Beyond the Quadruple Play: Networking, Convergence, and Customer Delivery

Comprehensive Report

IEC
Chicago, Illinois

Other Quality Publications from the International Engineering Consortium

- *Annual Review of Communications: Volume 60*
- *Business Models and Drivers for Next-Generation IMS Services*
- *The Basics of IPTV*
- *The Basics of Satellite Communications, Second Edition*
- *The Basics of Telecommunications, Fifth Edition*
- *The Basics of Voice over Internet Protocol*
- *Design and Test for Multiple Gbps Communication Devices and Systems*
- *Intellectual Property for Electronic Systems: An Essential Introduction*
- *IPTV Multimedia Networks: Concepts, Developments, and Design*
- *Mobile TV Research Report*
- *Moving Forward with 4G: Market Applications, Strategies, and Challenges*
- *VoIP and Enhanced IP Communications Services*

For more information on any of these titles, please contact the IEC publications department at +1-312-559-3730 (phone), +1-312-559-4111 (fax), *publications@iec.org*, or via our Web site (http://www.iec.org).

ISBN: 978-1-931695-61-9

Library of Congress Cataloging in Publications Data

Beyond the quadruple play: networking, convergence, and customer delivery.
    p. cm.
 ISBN-13: 978-1-931695-61-9
 1. Broadband communication systems. 2. Broadband communication equipment industry. 3. Convergence (Telecommunication) I. International Engineering Consortium.
 TK5103.4.B44 2007
 384--dc22

                                2007038970

International Engineering Consortium
300 West Adams Street, Suite 1210
Chicago, Illinois 60606-5114, USA
+1-312-559-4100 voice • +1-312-559-4111 fax
publications@iec.org • www.iec.org

About the International Engineering Consortium

The International Engineering Consortium (IEC) is a nonprofit organization dedicated to catalyzing technology and business progress worldwide in a range of high technology industries and their university communities. Since 1944, the IEC has provided high-quality educational opportunities for industry professionals, academics, and students. In conjunction with industry-leading companies, the IEC has developed an extensive, free on-line educational program. The IEC conducts industry-university programs that have a substantial impact on curricula. It also conducts research and develops publications, conferences, and technological exhibits that address major opportunities and challenges of the information age. More than 70 leading universities are IEC affiliates, and the IEC handles the affairs of the Electrical and Computer Engineering Department Heads Association and Eta Kappa Nu, the honor society for electrical and computer engineers. The IEC also manages the activities of the Enterprise Communications Consortium.

Table of Contents

Henrik Basilier, Expert, Multiaccess Edge, Ericsson
Jan Söderström, Director, Broadband and Transport Research, Ericsson
Howard Green, Senior Advisor, Broadband and Transport Research, Ericsson
Hans Mikelsson, Manager of Connectivity and Protocols, Broadband and Transport
    Research, Ericsson
Hans Byström, Director, Multimedia Applications, Ericsson
Ulf Jönsson, System Area Driver, Multiaccess Edge, Ericsson

Michael Boland, Senior Analyst, Interactive Local Media Practice, The Kelsey Group

Bill Bondy, Chief Technology Officer, Americas, Apertio

The Role of CPE to Provide Full Service Convergence
Peter Galyas, Chief Technology Officer, Tilgin
Dante Iacovoni, Marketing Director, Tilgin

The European Perspective – New Ideas from the Old Continent
Marcelo Garcia, Convergence Expert

A Shift in User Expectations
Natalie Giroux, Chief Scientist, Gridpoint Systems, Inc.

The Answer May Lie in Your Customer Service
Jeff Gordon, Senior Vice President, Innovation Center, Convergys

Ron Iannetta, Senior Manager, Communications and Media, BearingPoint

Heather Kirksey, Senior Manager, Standards and Emerging Technologies, Motive, Inc.

The IEC's University Program, which provides grants for full-time faculty members and their students to attend IEC Forums, is made possible through the generous contributions of its Corporate Members. For more information on Corporate Membership or the University Program, please call +1-312-559-4625 or send an e-mail to *cmp@iec.org*.

Based on knowledge gained at IEC Forums, professors create and update university courses and improve laboratories. Students directly benefit from these advances in university curricula. Since its inception in 1984, the University Program has enhanced the education of more than 500,000 students worldwide.

IEC Corporate Members

IEC–Affiliated Universities

The University of Arizona
Arizona State University
Auburn University
University of California at Berkeley
University of California, Davis
University of California, Santa Barbara
Carnegie Mellon University
Case Western Reserve University
Clemson University
University of Colorado at Boulder
Columbia University
Cornell University
Drexel University
École Nationale Supérieure des Télécommunications de Bretagne
École Nationale Supérieure des Télécommunications de Paris
École Supérieure d'Électricité
University of Edinburgh
University of Florida
Georgia Institute of Technology

University of Glasgow
Howard University
Illinois Institute of Technology
University of Illinois at Chicago
University of Illinois at Urbana-Champaign
Imperial College of Science, Technology and Medicine
Institut National Polytechnique de Grenoble
Instituto Tecnológico y de Estudios Superiores de Monterrey
Iowa State University
KAIST
The University of Kansas
University of Kentucky
Lehigh University
University College London
Marquette University
University of Maryland at College Park
Massachusetts Institute of Technology
University of Massachusetts

McGill University
Michigan State University
The University of Michigan
University of Minnesota
Mississippi State University
The University of Mississippi
University of Missouri-Columbia
University of Missouri-Rolla
Technische Universität München
Universidad Nacional Autónoma de México
North Carolina State University at Raleigh
Northwestern University
University of Notre Dame
The Ohio State University
Oklahoma State University
The University of Oklahoma
Oregon State University
Université d'Ottawa
The Pennsylvania State University

University of Pennsylvania
University of Pittsburgh
Polytechnic University
Purdue University
The Queen's University of Belfast
Rensselaer Polytechnic Institute
University of Southampton
University of Southern California
Stanford University
Syracuse University
University of Tennessee, Knoxville
Texas A&M University
The University of Texas at Austin
University of Toronto
VA Polytechnic Institute and State University
University of Virginia
University of Washington
University of Wisconsin-Madison
Worcester Polytechnic Institute

Table of Contents by Author

Full-Service Broadband Architecture

Henrik Basilier
Expert, Multiaccess Edge
Ericsson

Jan Söderström
Director, Broadband and Transport Research
Ericsson

Howard Green
Senior Advisor, Broadband and Transport Research
Ericsson

Hans Mikelsson
Manager of Connectivity and Protocols, Broadband and Transport Research
Ericsson

Hans Byström
Director, Multimedia Applications
Ericsson

Ulf Jönsson
System Area Driver, Multiaccess Edge
Ericsson

Executive Summary

After years of talk about fixed-mobile convergence (FMC) and next-generation networks (NGNs), technology solutions are now ready to give fixed and mobile operators a major leap forward. Full-service broadband provides people with simple and convenient access to all their communications, entertainment, personal media, and Internet services, wherever they are and from all of their connected devices. Operators have an opportunity to deploy an open, standards-based, combined fixed and mobile architecture that offers a cost-effective, evolutionary route to full-service broadband.

For users, convenience is a key success factor. Converged services need to be seamlessly and intuitively accessible across all devices and networks. For operators, the introduction of new services and additional network capacity must incur minimal additional cost of ownership.

The full-service broadband architecture is designed to meet these needs across residential and enterprise service offerings. At its core is a reliable, secure, and cost-optimized transport network.

Overlaid on this are a variety of access technologies, each evolving in support of full-service broadband, with access heterogeneity handled through multiaccess edge capabilities. Internet protocol multimedia subsystem (IMS) is a key enabler for control and delivery of user convenient end-to-end services, reachable from any device and across any access technology. User mobility and multiaccess connectivity are enabled through consistent and open user-to-network interfaces (UNIs), while open network-to-network interfaces (NNIs) ensure interoperability with partners such as other operators and enterprises.

Such an evolved, open-standard architecture is essential to build a profitable and sustainable full-service broadband business. It provides the consumer electronics industry with the required economies of scale. It drives usage by offering the user transparency and convenience, enabling anyone to reach anybody (or any device) at any time, and making the same services accessible anywhere. It improves cost-efficiency by stimulating competition and simplifying interoperability and management. Above all, it encourages a common ecosystem that is beneficial to all parties involved.

Key Challenges and Drivers

Capitalizing on Full-Service Broadband Connectivity

Users will increasingly expect all their services to be accessible anywhere and from any device. They want to be constantly connected and able to communicate with friends, family, and coworkers. While working, they need to be connected to their enterprise network environment to access e-mail and files. Outside work, they want to be connected to residential networks to access their personal media collections and other content.

Today's multi-play approach does not provide the full answer to meeting these user needs. Triple play leaves the mobile broadband service revenue up for grabs by competitors, while a basic bundled fixed and mobile subscription offering does not provide sufficient integration to reach the full potential of full-service broadband in terms of revenue growth and reduced churn.

What is needed is an architecture that enables seamless connectivity across fixed and mobile access boundaries. This architecture needs to deliver broadband connectivity and standardized multimedia services to a wide range of devices, including media servers, video cameras, portable media players, personal computers (PCs), and mobile phones. The architecture must be able to satisfy the consumer electronics industry's need to achieve economies of scale while providing smooth integration with enterprise environments. It must deliver a solution based on open standards that is acceptable to all parties and that cultivates a common ecosystem—just like the mobile industry today.

Ubiquitous availability of any service increases its value to users and, therefore, users' willingness to pay a premium. The value of communications and networking capabilities also increases with the number of connected users able to share the service. The success of the open standards-based Global System for Mobile Communications (GSM) network is proof of these two tenets.

In the new full-service broadband architecture, basic IP/Ethernet connectivity needs to be cost-efficient, secure, and reliable. This is particularly important not only for enterprise services, but also for consumer services such as TV.

To capture the revenue potential and the competitive edge associated with broadband services, the scope of broadband connectivity needs to be extended to mobile as well as fixed access. New quality-assurance measures are also needed to deliver a mix of Internet, IP virtual private networks (IP–VPNs), and IMS–enabled services. The new architecture must provide a sound foundation for both retail and wholesale broadband service offerings and must support user mobility with consistent capabilities that enable connectivity services to be charged for.

The Importance of User Convenience

User convenience is fundamental to drive the mass market of broadband services, especially as the variety of devices and services grow. User convenience encompasses security, simplicity, personalization, and look and feel. It is about always being able to communicate and reach services in a consistent and intuitive way, independent of which access is being used. Likewise, the ability to connect a device to a wireless or wired premises network or to a mobile network in a simple, convenient way is key to enhancing user satisfaction and minimizing interactions with support centers. The value of such factors grows with the number of devices supported.

Standard network technologies such as Ethernet, Wi-Fi, and wideband code division multiple access (WCDMA)/high-speed packet access (HSPA) are increasingly being integrated into devices. Open standards should boost user convenience further by making service activation uniform and service usage seamless and simple.

Managing Cost of Ownership

The full-service broadband architecture needs to support services that are both affordable for users and profitable for operators.

The growth of enterprise and video traffic—for TV services and user-generated video—over fixed and mobile networks is driving demand for higher network capacity. The move to interactive and personalized high-definition TV (IPTV/HDTV) services will require high-throughput deep-fiber access networks. Higher bandwidth requirements are driving access technology evolution, for example, through very-high–data-rate digital subscriber line 2 (VDSL2), gigabit passive optical networks (GPONs), and third generation (3G) long-term evolution (LTE).

The transport networks underlying full-service broadband services face significant challenges in the form of huge increases in scale, massive new bandwidth demands, and the need to reduce the cost per transported bit. This means savings need to be found in network deployment and operation. Moreover, there is uncertainty about the size, shape, and timing of these new requirements, primarily because of TV. Therefore, it is crucial to maintain network flexibility to be able to respond to changes.

Telecom management is another key factor in managing cost of ownership, especially in deployment and provisioning. In the full-service broadband architecture, a particular challenge is that some functions are access-specific, while some are common.

The full-service broadband architecture must be designed to minimize the cost of introducing new services and new access technologies. It must offer the flexibility to allow for the incremental addition of services and accesses. Basing it on open standards will drive volumes and economy of scale, enable interoperability, and reduce operators' cost of ownership.

The Full-Service Broadband Architecture

Architecture Overview
The basic full-service broadband architecture, as shown in *Figure 1*, delivers a consistent, multilayer UNI. This is a prerequisite for users to be able to access any broadband service anywhere from any device and for operators to achieve the desired economies of scale. The NNIs and roaming interfaces drive usage by enabling anyone to reach any other device, network, or service anywhere.

To allow service providers to offer easy-to-use services, it is essential that functions which are common between several services are aligned and that serv-

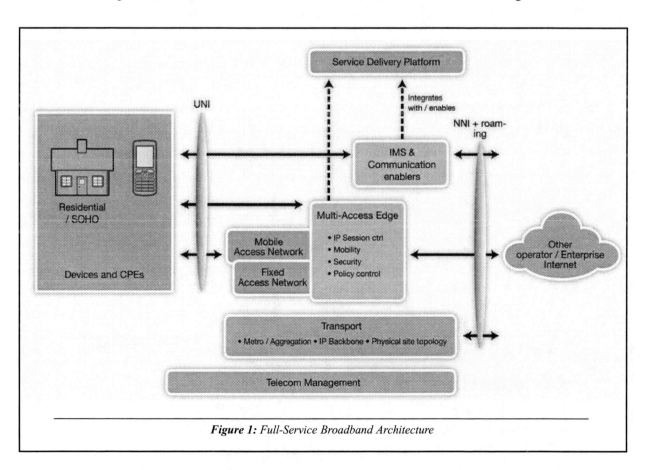

Figure 1: Full-Service Broadband Architecture

ices work well together. This is the foundation for the IMS and communication enablers layer (see [12]). It offers standard services such as multimedia telephony, messaging, and presence. It also provides interfaces to be used when developing more customized applications such as IPTV.

On top of IMS and communication enablers resides the operator service delivery platform (SDP, see [1]). It contains the operator business processes for managing and selling services. The SDP also connects these business processes to customer-relationship management, billing, network, and other essential supply chain elements such as residential and enterprise self-service portals.

Access networks implement technology-specific functions—for example, in the radio base station (RBS) or DSL access multiplexer (DSLAM). The evolution of access technologies enables and drives the evolution of services. The multiaccess edge domain contains access-specific as well as common capabilities for mobility, authentication, roaming, policy control, and charging. Third-generation partnership project (3GPP) policy and charging control (PCC) and system architecture evolution (SAE)

provide the foundation for integrating 3GPP and non–3GPP accesses.

The common transport layer is designed for carrier-class, cost- and performance-optimized routing and switching. It is kept independent of service delivery capabilities so that these can be independently introduced, scaled, and distributed. The transport layer provides transmission and IP–VPN services to higher-layer services and external networks, e.g., for enterprises. The routing and switching are access technology-independent, enabling different access technologies to co-exist and be upgraded or introduced easily.

The physical location of functionality depends on several factors, including site topology and costs, geographical distribution of users, service utilization transport costs versus site operational expenses (OPEX) and capital expenses (CAPEX), level of mobility, and the location of peering points. All in all, this calls for flexibility so that functions can be optimally distributed using open Internet protocols. The need for flexibility becomes even more evident when considering how to map functionality onto the topological site infrastructure, as illustrated by

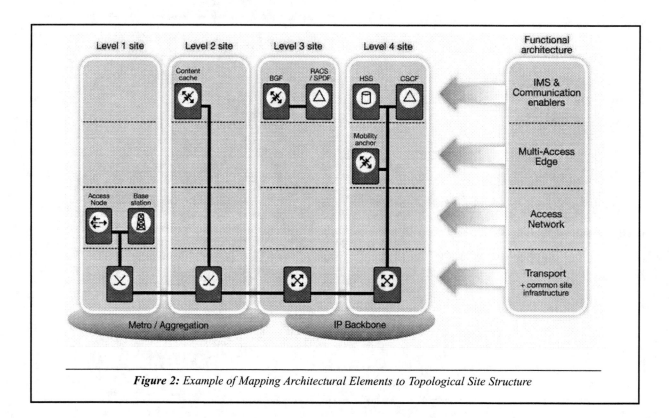

Figure 2: *Example of Mapping Architectural Elements to Topological Site Structure*

the example in *Figure 2*. For guidance on how distributed co-sited functionality can be managed and how migration of functionality between topological levels is enabled, see [2].

Each area needs to be managed and therefore ties into the overall telecom management structure (see [3]).

Building a Transport Infrastructure for Flexibility

The aggregation and transport parts of the full-service broadband architecture are designed to be common to all fixed and mobile access technologies. They are designed for low cost of ownership and based on carrier-grade Ethernet. Common transport is aided by the fact that services are becoming increasingly homogeneous, independent of access type: the main payload will be IP packets carried by Ethernet frames.

Ethernet is the uniting link-level protocol, over both native Ethernet (IEEE 802.3) links, or in combinations with media-specific transport technologies such as Ethernet over microwave and Ethernet over optical transport. Carrier-grade Ethernet is about adding components—primarily software—for handling of scalability, fault location, and performance monitoring.

Transport network aggregation is flexible in terms of packet forwarding and traffic separation and it supports Ethernet bridging, tunneling techniques such as multiprotocol label switching (MPLS), and the use of IP routing protocols. The connectivity services provided by the aggregation network seem likely to converge on the Metro Ethernet Forum definitions of Ethernet services such as E-Line, E–LAN, E-Tree, and their variants. These are increasingly being used internally in the operators' networks and for enterprise connectivity services. Also mobile networks will adopt these standards and migrate from a world of predominantly time division multiplexing (TDM) leased circuits. For more on metro aggregation, see [13].

Fiber access networks such as point-to-point Ethernet and different flavors of PON are being built out to support advanced media delivery (e.g., TV/HDTV). In combination with the buildout of remote VDSL2 sites, this creates a deep-fiber

aggregation architecture that is one of the cornerstones of the full-service broadband aggregation and transport architecture. There is a trend—particularly in urban areas—of aggregation networks going from several aggregation stages toward a flatter optical network with fewer active nodes. This is enabled by long-range optical transceivers in combination with energy-efficient, high-capacity, passive C/DWDM fiber multiplexing technology. The flattening of the aggregation network in turn opens up opportunities for cost reduction in network operation and real estate consolidation. The separation of transport functions from service and access enables such consolidation to be carried out incrementally.

In rural areas or developing countries, the fiber trend will be weaker and multilevel aggregation topologies will still be the preferred solution. Here, microwave technologies will also have a crucial role to play in efficient aggregation and transport of mobile and fixed broadband traffic.

Transport network OPEX is reduced through a unified management view. More automated procedures are introduced, for example, for flexible sharing of capacity between multiple business roles. This includes auto-discovery of equipment and connectivity and monitoring and reporting of service-level agreements (SLAs).

Connecting Devices and Users

For residential and small office/home office (SOHO) access, full-service broadband demands a shift to deliver services to individual users rather than to a household, as depicted in *Figure 3*.

Users increasingly want to do more on the move, and they use a plethora of devices, including portable media players, gaming devices, and set-top boxes (STBs). They also want to get secure and transparent access to e-mail, faxes, and servers. All services should be available to them, wherever they are, from any device, and independent of access. Likewise, user devices need to be connected to other devices in the residential network, wherever they are—for example, for remote access to personal media. Emerging standards such as the Digital Living Networking Alliance (DLNA) will play a central role here.

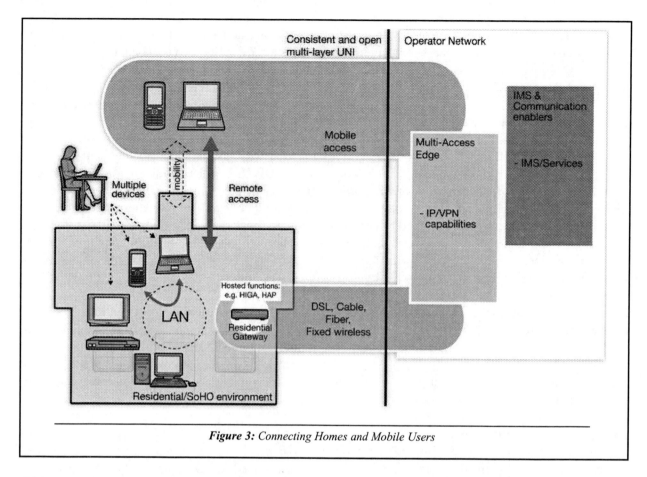

Figure 3: *Connecting Homes and Mobile Users*

With a consistent and open UNI, service usage is made convenient for users. From a consumer electronics industry perspective, this uniformity delivers economies of scale. IMS is a cornerstone for this UNI, along with standard IP/Ethernet and VPN capabilities. For multiaccess mobility, 3GPP SAE will provide the baseline.

The ability to provide services to residential or SOHO networks—and to the consumer electronics devices connected to them—is crucial to the success of full-service broadband. The full-service broadband architecture takes a pragmatic approach that fits a variety of operator strategies. For example, it supports interoperability between IMS services and non–IMS–capable consumer electronics devices through intelligent home gateways. It also includes operator-provided home access points (HAPs). To counter the risk of the escalating cost of ownership, the architecture also enables a more network-centric approach, where the operator-provided residential network functionality is kept to a minimum.

The full-service broadband architecture is designed to make the user experience transparent and painless—for example, by avoiding the need for hands-on configuration, with the operator as an enabler.

Evolving Access Networks for Full-Service Broadband

Full-service broadband services—especially video services—are enabled by and drive advances in access technologies, both in terms of increased bandwidth and reduced packet delays. The most important examples are VDSL2 (see [4]), fiber-to-the-home (FTTH, see [15]) technologies, WCDMA/HSPA (see [14]), and 3G LTE (see [5]). Wireless access will be used both for mobile and stationary fixed-wireless access.

Although these technologies will be competing with each other, it is also increasingly important that they can be used in a complementary way in a single solution to satisfy different economic needs. In some cases, FTTH may be the most cost-effective solution, especially where the cost of fiber deployment is comparatively low. In other cases, VDSL2 will make the most economic sense. If the copper loops cannot be made short enough in an economically justifiable way, the use of wireless access such as WCDMA/HSPA or LTE may be most cost-optimal. Wireless access also offers the added value of mobility.

If the same services are to be provided across multiaccess technologies, the delivered capabilities—especially maximum bit rate—need to be comparable across these accesses. A competitive mobile broadband solution becomes a necessity for a successful full-service broadband business. With HSPA evolved and LTE as the underlying access technologies, mobile broadband supports the high bit rates, latency, capacity, and low cost per delivered bit needed for full-service broadband. In addition, mobile broadband provides wide coverage and full mobility, making full-service broadband access available everywhere and at any time.

The advances in access technology also open new opportunities when used in combination. For instance, an existing GPON solution can be used to backhaul traffic from RBSs and DSLAMs.

The full-service broadband architecture is designed to simplify the co-existence of access technologies and changes such as the addition of new technology. This is accomplished through access-agnostic transport layer and service delivery capabilities.

The Importance of the Multiaccess Edge

The access heterogeneity of full-service broadband must be managed to give users the best possible convenient connectivity with low cost of ownership for operators. The multiaccess edge capabilities, shown in *Figure 4*, have a key role to play here.

At the core of multiaccess edge are access- and application-independent capabilities for subscriber management, policy control, resource management, deep packet inspection, security, media gating, real-time charging, and mobility management. A multiaccess edge also provides access interworking functions to interface both wireless and wireline access networks. A service interface ensures that access

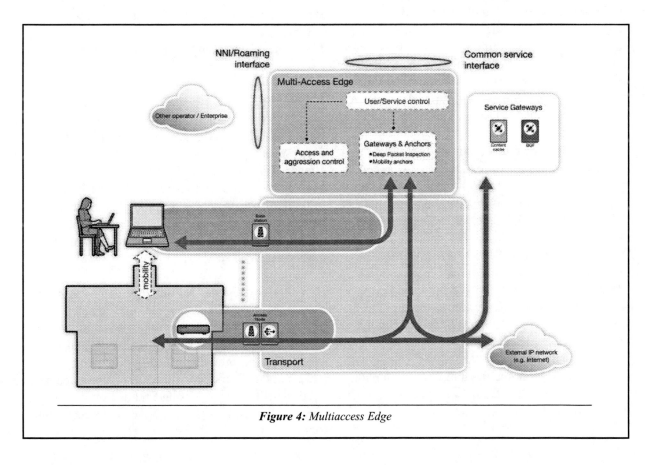

Figure 4: Multiaccess Edge

dependencies are hidden from higher-level services such as IMS.

Multiaccess edge capabilities are often bundled with edge routing and/or aggregation/switching capabilities, as this is the optimal placement of this control. It may also bundle functions supporting the applications, e.g., content caching.

The data plane functions are typically implemented in multiservice edge routers (MSERs), which can be deployed in single or multi-edge configurations. The control plane is realized by policy controllers. Multiple standards exist in this space: 3GPP (gateway general support node [GGSN], plesiochronous digital hierarchy gateway [PDN GW], policy and charging rules function [PCRF]), DSL Forum (broadband network gateway [BNG]) and Telecoms and Internet Converged Services and Protocols for Advanced Networks (TISPAN) (IP edge, resource and admission control subsystem [RACS]). Convergence/alignment between these is an important component of full-service broadband.

Both nomadic mobility and full mobility is supported. For user convenience and cost control, a consistent approach to authentication is achieved through credentials that are tied to the user—such as a subscriber identity module (SIM) card—and that are used across access technologies. Roaming and other forms of business interworking are also key. All these capabilities are based on 3GPP SAE (see [7]).

Access and aggregation resources such as admission control are, to a large extent, access-dependent and need to be managed. For fixed access, TISPAN RACS (see [8]) is the baseline, while mobile access will use inherent 3GPP capabilities.

IMS as an Enabler of Standardized Multimedia Services

IMS provides an open and powerful way to provide standardized networking and communications between any device and any access technology. Standardized functions comprise presence and group management for easy networking, along with standardized services such as multimedia telephony (MMtel), which will eventually replace all circuit-switched fixed and mobile voice.

IMS also provides simple application programming interfaces (APIs) so that these networking and communications tools can be easily integrated by developers and service providers into their own applications, be it gaming or Web 2.0 networking or collaboration services. This is essential to allow operators to differentiate their service offering and also for them to form beneficial relationships with potential Internet service partners. The use of a Java-based API for the client ensures the basic IMS capabilities and the applications that sit on top of them can be easily integrated into a wide range of devices.

Open and standardized UNIs and NNIs are prerequisites for communication services between endpoints (devices or servers). The UNI provides a common way to access services across devices and networks. This ensures volume availability of devices and drives usage by making the services available anywhere. The NNI enables the services to work between networks, operators, and countries; it also supports the business arrangements for interconnection and termination charges typical for communication services. Together these standardized interfaces are what ensure that IMS communications can work between any devices on any IP–based network, anywhere in the world.

IMS has been adopted as the common framework for building a variety of communication services based on these principles. IMS also provides end-to-end quality of service (QoS) control and, for this purpose, integrates with the multiaccess edge and control capabilities. The common IMS standard enables communication services that are built on it to be used across access technologies. An example is MMTel for multimedia telephony services, used for both mobile and fixed access. For more information on IMS, see [10].

IMS is a key enabler for delivery of user convenience across access technologies and seamless mixing of services. With the full-service broadband architecture, users can access their personal IPTV services—e.g., broadcast, video on demand (VoD), and a network personal video recorder—independent of whether they are using a mobile device on the train or watching a flat-screen TV in the living room. Also, the IPTV services can be seamlessly combined with other services such as IMS–based chat or messaging, presence and buddy lists, and other Internet-based services. The personalization of IPTV services and their combination with communication services paves the way for interactivity in areas such as advertising and game shows. For more details on IMS–based IPTV, see [11].

The use of IMS functions and interfaces for billing and provisioning cuts the time and costs involved in introducing new services such as IPTV and helps reduce OPEX. IPTV benefits from the generic IMS mechanisms of bandwidth negotiation and end-to-end QoS control. This provides a simple means to optimize quality for IPTV sessions.

Recommendations and Conclusion

The full-service broadband architecture—in which all services are made available anywhere, anytime, and to any device—represents a tremendous opportunity for the industry. This architecture can be built using components that are already, or soon will be, available based on open standards.

The recommendation for deploying transport networks is to balance the short-term needs of optimizing fixed broadband networks with the flexibility required for full-service broadband. It must be simple and cost-efficient to add multiaccess mobility, new access technologies, and new services. The transport network needs to be flexible, yet optimized for secure, reliable, and cost-efficient packet delivery.

Services should be built in such a way that they can be made available to mobile as well as fixed users. It is important to recognize that full-service broadband, unlike fixed broadband, is about providing individualized services to users, not to a particular residence.

It is worth preparing for changes in topology and site structure. As bandwidth demand grows and fiber is deployed deeper into the network, the economics will change. It should be possible to move higher-layer functions such as multiaccess edge and IMS over time.

Adding uniform multiaccess capabilities will enable convenient connectivity for users. It must be possible to upgrade/add accesses and to extend the reach of services through business-to-business arrangements in areas such as roaming. Therefore, network deployment should move toward standards that are driven in the direction of full-service broadband.

The ability to reach anyone or any service, from anywhere, using any device will grow end-user consumption. It will stimulate competition and simplify operation, thus driving cost-efficiency. Full-service broadband will provide a profitable growing business for the operator and all parties involved.

References

[1] Service Delivery Platform-Efficient Delivery of Services, Ericsson, www.ericsson.com/technology/whitepapers.

[2] Ericsson's Integrated Site Concept, Ericsson Review No. 01, 2005, www.ericsson.com/ericsson/corpinfo/publications/review.

[3] Enhancing Telecom Management, Ericsson, www.ericsson.com/technology/whitepapers.

[4] VDSL2: Next Important Broadband Technology, Ericsson Review No. 01, 2006, www.ericsson.com/ericsson/corpinfo/publications/review.

[5] The long-term evolution of 3G, Ericsson Review No. 02, 2005, www.ericsson.com/ericsson/corpinfo/publications/review.

[6] Evolution of policy control and charging, 3GPP TS 23.203.

[7] 3GPP system architecture evolution: Report on technical options and conclusions, 3GPP TR 23.882.

[8] TISPAN ES 282003: "Resource and Admission Control Subsystem (RACS)."

[9] DSL Forum: TR-101 Migration to Ethernet Based DSL Aggregation, April 2006.

[10] IMS – IP Multimedia Subsystem, Ericsson, www.ericsson.com/technology/whitepapers.

[11] Evolving the TV Experience, Anytime, anywhere, any device, Ericsson Review, No. 03, 2006, www.ericsson.com/ericsson/corpinfo/publications/review.

[12] Services in the IMS ecosystem, Ericsson, www.ericsson.com/technology/whitepapers.

[13] Full Service Broadband Metro Architecture, June 2007, www.ericsson.com/technology/whitepapers.

[14] HSPA, the Undisputed Choice for Mobile Broadband, May 2007, www.ericsson.com/technology/whitepapers.

[15] Deep-fiber broadband access networks, Ericsson Review, no. 01, 2007, www.ericsson.com/ericsson/corpinfo/publications/review/2007_01/01.shtml.

Triple and Quad Play: Who Will Win the Bundled Service Battle?

Michael Boland

Senior Analyst, Interactive Local Media Practice
The Kelsey Group

Summary

Telecom companies are collectively spending about U.S. $50 billion to roll out the infrastructure and technologies that will drive the next generation of voice, data, and video services. These bundled services, known as "triple play," represent the basis for telecom product strategies over the next decade. Adding wireless to the package brings in a fourth dimension, thus the term "quadruple play." It will not be enough for telecom companies to bring in revenues from bundled services to recoup their massive investments, even in light of the associated retention benefits. The opportunity for telecoms will lie in positioning themselves as both data networks and providers of aggregated content, and targeted Internet protocol (IP)–based advertising to stationary and mobile devices will be a big part of this. Cable companies will also continue to play a role by adding voice over IP (VoIP)–based phone service to existing video and data services and forming partnerships to provide wireless service. Telecoms are currently fighting for the legal rights to franchise these services and lay down the massive fiber networks they require. But the real bundled service battle will be in developing and marketing the continuity of service and content across different devices (e.g., television, computer, telephone, wireless models). The fate of the "net neutrality" debate, content aggregation, technology development, and incumbent positioning in these service areas will underpin the competitive advantages that telecoms and cable companies each will hold.

Triple Play Heats Up

On September 29, 2006 California Gov. Arnold Schwarzenegger signed into law a bill that will alter the landscape of phone video and Internet services as we know it. The legislation, which took effect January 1, 2007, allows telephone companies to obtain statewide licenses in California to offer video services. Without this, telecommunication companies such as AT&T and Verizon would face the daunting task of obtaining licenses in each California municipality.

The tens of millions of dollars these companies collectively spent to lobby for the bill (independent of their massive infrastructure deployment costs) is telling of its importance to them. Indeed, video will be a key component of the triple- and quad-play bundled services that will be telecoms' biggest weapons against cable companies, which are moving in on their traditional voice businesses.

Telecoms' move into broadband-delivered video, or IPTV, is at first glance a countershot to protect territory from cable companies. Cable companies, meanwhile, see a threat in telecom IPTV offerings and have accordingly initiated a strong marketing push for their VoIP services. This self-propagating battle has begun to heat up in areas of the country where telecoms have gained a statewide right to provide TV service, including Texas and New York. Next up is California.

This will result in decreased prices for service bundles that include voice, video, and broadband—a marketing strategy that offers one statement with a billing amount that is less than the sum of its parts. The increased choice in the market has led consumer advocates to support telecoms' rights to seek statewide franchise agreements, as additional competition increases choice and drives down price.

The Inside Track

Many factors should be examined to see which company is better positioned to win the bundled service battle. This comes down to matters of technology, incumbent positioning, organizational execution, and marketing. It is clear that cable companies are the incumbents when it comes to video, while telecoms hold that position in voice service. Cable companies, however, are at an advantage because it is easier and less expensive to establish and roll out a voice service than an IPTV offering.

However, this battle could also be waged in a fourth dimension in which telecoms currently have the upper hand: wireless. Wireless service and mobile search will be significant revenue generators for telecoms and will allow them increased economies of scale with billing, infrastructure, and cross-platform ad sales that could involve location-based services in addition to targeted ad delivery across other packages. The targeting enabled by IPTV and mobile search together will also present interesting opportunities for telecom-owned directory operations.

"It's a lot easier to deliver voice than it is to deliver video, so you have to give the first-run advantage to cable. Over the long haul though, the telecoms have a shot at a quad play given their stake in wireless," said John Reed of Bluestreak Network during a panel at Dow Jones' Network Ventures conference in February. Reed is chairman of Bluestreak, a Dallas-based mobile and digital TV technology provider. "They are in a very strong long-term position if they can pull all the way through to wireless. They just have to get this video piece figured out, and they have really been struggling to get out of the gate with a consistent platform that is best of breed. As soon as they find that combination, they will be in a strong position."

Cable companies have begun to seek partnerships to acquire wireless capability to remain competitive as the battle evolves from triple to quad play. Comcast, Cox Communications, and Time Warner Cable, for example, have entered agreements with Sprint Nextel to provide cable customers with co-branded wireless services.

In the meantime, cable's "first-run advantage" can already be seen in the numbers. Cable operators ended the second quarter of 2006 with more than 6 million voice customers, a net gain of 677,000 over the previous quarter. Telecoms conversely had 2 million video subscribers at the end of the quarter. Assuming a zero-sum game for the sake of argument, cable companies over time have netted roughly three times as many telecom voice customers as telecoms have taken cable video customers.

Related to this is the concept that relatively few consumers are inclined to switch to IPTV when the initial service rollouts will be on par with current cable offerings.

"In a broader context [IPTV] is hard to deliver. It will be interesting to see how quickly and compelling the telecom rollouts will be because they are spending a lot of money and trying to line up all of the content rights themselves, subsidizing set-top boxes and setting up your home network," said Menlo Ventures' Shawn Carolan at the Dow Jones event. Carolan is managing director of the Menlo Park, California–based venture capital fund. "When you look at what they are spending to try to give most of these services in parity with cable and satellite, it will be interesting to see how many consumers actually switch over that quickly. I'm somewhat pessimistic that it will happen in the U.S."

In fact, a poll conducted in March by Research and Markets showed only 5 percent of satellite or cable TV subscribers would "definitely switch" to a telecom IPTV. Thirteen percent said they would "probably switch," and 52 percent were undecided.

Though the huge investments in fiber infrastructures that telecoms are making will represent a foundation for intriguing IPTV offerings superior to current cable service features, the initial packages will be on par with cable, as suggested by Carolan and the Research and Markets survey results. This comes down in part to a matter of content aggregation challenges and the dearth of content they will cause for initial IPTV service packages. In other words, there will be a disconnect between IPTV's content capacity advantages and its content aggregation disadvantages (having to start from scratch). As a result, telecoms are rapidly pursuing deals to

fill up their massive capacity with content that will attract subscribers.

The technical advantages of IPTV will indeed allow for a great deal of content. This is because of its IP delivery and switched video architecture, which is similar to the way Web pages are delivered to computers.

Cable Telephony Subscriber Counts (000)	2Q 2005	3Q 2005	4Q 2005	1Q 2006	2Q 2006
Comcast*	1,228	1,230	1,242	1,321	1,463
Cox Communications**	1,506	1,607	1,697	1,807	1,892
Cablevision***	487	609	739	873	994
Charter Communications	68	90	122	191	258
Insight	74	81	90	100	107
Mediacom Communications	NA	2	22	46	66
Time Warner Cable	614	854	1,100	1,370	1,604
Total	3,976	4,474	5,012	5,708	6,384
Quarterly Adds.	(3,467)	498	538	696	677
Total Telephony Homes Passed by Group	44,641	50,766	59,321	69,727	78,030
Subs. Percent of Telephony Homes Passed	*8.9%*	*8.8%*	*8.4%*	*8.2%*	*8.2%*

*At the end of Q2 2006, Comcast passed 25.6 million homes with VoIP service and counted 721,000 VoIP or digital voice customers. The remaining homes passed and subscribers reflected legacy circuit-switched.

**EMDI estimates. Cox offers a mixed of circuit-switched and VoIP service.

***Includes sequentially decreasing, and increasingly trivial, amounts of legacy circuit-switched customers.

Source: IP Democracy Forum (2006)

Incumbent Telco Video Subscribers (000s)					
Telco	2Q 2005	3Q 2005	4Q 2005	1Q 2006	2Q 2006
AT&T	404	419	457	491	533
BellSouth	394	460	523	628	691
Qwest Communications	120	151	183	228	273
Verizon*	250	305	349	415	485
Total	1,168	1,335	1,512	1,762	1,982
Net Change	*161*	*166*	*178*	*249*	*220*

*Q2 2005 is an estimate.

Source: IP Democracy Forum (2006)

Table 1: *Cable Telephony versus Telecom Video Subscribers*

"[The] switched IP platform gives us the capacity and flexibility to offer increased programming choices, including a wide array of content that competitors may not be able to offer due to spectrum constraints," an AT&T spokesperson told The Kelsey Group. "[It] creates the ability to offer more content, including more niche content, than is possible with a traditional broadcast cable model. And features like the ability to search for programs and actors using a keyword search and having video-on-demand [VoD] titles incorporated into the channel lineup improve the customer access to on-demand content."

In the same way the Web has propagated the economic principle of the long tail, IPTV will do so with video content. This will also change the volume of inventory targeting capabilities and economies of advertising, which will have local advertising implications.

	Verizon	Cablevision	Notes
Expanded basic service	U.S. $39.95	U.S. $46.95	Verizon costs US $34.95 with one-year agreement. Cablevision does not offer yearly contracts.
Standard-definition set-top box	U.S. $3.95	U.S. $5.74	—
Number of standard-definition channels on expanded basic	148	92	Includes local stations, public access, and regional sports.
Number of national standard-definition channels on expanded basic	125	50	Not including broadcast network affiliates or regional sports channels.
Additional TV channels on digital tier	NA	36	Cablevision's iO Digital costs U.S. $9.95.
HDTV set-top box	U.S. $9.95	U.S. $5.74	—
Minimum additional payment for HDTV	None	U.S. $4.95	Cablevision's iO Navigation package does not include the digital-tier package of standard-definition channels.
Total number of non-premium HDTV channels	18	16	Includes local stations and regional sports.
Number of premium national HDTV channels available	5	5	—
Additional payment for HBO	U.S. $14.95	U.S. $11.95	Both include multiplexed versions of HBO.
HDTV box with a DVR	U.S. $12.95	U.S. $15.69	Cablevision charges U.S. $5.74 for the box and U.S. $9.95 for DVR service.
* — not including special discounts or bundles			Sources: Verizon, Cablevision Systems

Source: USA Today (2006)

Table 2: *Side-by-Side Comparison of Cable and IPTV Service in Hempstead, New York*

"IPTV has the power to create a new, more effective model for advertisers and to enhance the relevance of advertising messages for both consumers and advertisers," according to AT&T. "We believe these new capabilities will have tremendous appeal for advertisers, programmers, and consumers."

This attractive vision of the future for telecoms will take a long time to manifest, however, because of the content aggregation challenges and slow, costly infrastructure deployments they currently face (not to mention the legal hurdles mentioned above, although we do not see these ultimately standing in the way).

Telecoms bring hardware and software technologies to IPTV delivery that have not been used in unison on a large scale (e.g., new fiber networking equipment, IPTV set-top box software, back-end server software). Currently the largest IPTV deployment in the world is in Hong Kong, where broadband carrier PCCW has 650,000 subscribers. That is about half the population of San Antonio, which is AT&T's first market (and home to its headquarters). It remains unclear how the many disparate technologies will stand up to the strain of millions of users simultaneously accessing the system.

But once these issues are hammered out and IPTV reaches its potential, it will have technical abilities far superior to what cable television's current architecture allows. The good news for cable companies is that this will not happen for another three to five years, meaning they have that much time to enhance their own levels of service to compete. Comcast, for example, recently told *The Wall Street Journal* it plans to develop extensive viewing guides that aggregate its TV schedule with on-line video clips and VoD options that give customers a portal for entertainment and shopping, akin to the Web itself. Until then, service bundling will be an important tool for customer retention, price competition, and continued growth in voice users.

The Long-Range Battle

Gaining customers with compelling video services and competing on price will constitute the first phase of the bundled service competition. This will determine the subscriber bases each provider holds,

the true monetization potential of which will come during the next step: serving personalized on-demand content and targeted search-based advertising across devices. Continuity of service will also be a key tenet of bundled packages, which will serve retention and marketing purposes and also allow for acute behavioral, geographic, and contextual targeting.

Yahoo had this concept in mind when it launched Yahoo Go at the annual Consumer Electronics Show in January. The product enables users to access personalized content on any connected device. Instead of storing data separately on individual devices, all a user's content resides on Yahoo servers. This means changes to data or personalization features on, say, a cell phone are also reflected on the user's PC, television, and other connected devices. This is one vision of bundled services' future.

Terry Semel, Yahoo's chairman and chief executive officer, also gave a glimpse into the advertising opportunities that come with this continuity of service and content, including news, instant messaging (IM), e-mail, photos, music, and video. This not only can "hook" users with Yahoo products that are cemented in the interfaces of myriad devices, but also can have users' data, preferences, and search history all in one place, opening up behavioral targeting opportunities. Combine that with the geotargeting capabilities of mobile devices (global positioning system [GPS] will be ubiquitous in mobile phones by the end of 2007, as we have previously noted) and with the IP targeting capabilities of IPTV, and you can begin to picture the possibilities for service providers.

Barriers to Entry

The challenge for telecoms in the utopian scenario outlined above is they do not own the content or platforms that represent the "killer apps" of these communication media. They are rather the medium through which content flows—in other words, a "dumb pipe." To unleash the true potential of bundled services, this has to change.

Telecom providers know this and have begun to travel down the long road of content aggregation and Hollywood deal-making for which there is a steep learning curve. Spurring this on has been telecoms' realization of the lead time required to line up content to fill the vast reaches of their IPTV service packages. The massive investment in high-bandwidth fiber network deployment will, they realize, be for naught if there is no content in place to lure users away from cable.

Telecoms are quickly beginning to realize that this is a costly and time-consuming effort for which their lack of experience could be disabling (this is another place where cable companies have an advantage). Further, the distribution platforms and social media software that will dress these service packages (e.g., IM, e-mail, music, video) also represent a gaping hole.

To gain this access, experience, and expertise, telecoms will have to form partnerships with or acquire tech companies. This and only this will allow them to reach the potential of bundled services.

Battle Lines Being Drawn

When we talk about the telecoms that will be mired in the bundled service race, it is more accurate to simply name AT&T and Verizon, as recent industry consolidation has extended their combined footprint throughout most of the United States. Similarly, a few cable giants cover nearly the entire country, most notably Comcast, Cox, and Time Warner Cable. Each individual market will therefore come down to a competition between one major telecom and one cable company.

The relationships telecoms have already formed have clarified to some degree the fate of their content and platform partnerships. Yahoo and AT&T have a long-standing arrangement (via SBC and BellSouth) in which Yahoo offers free content and services to AT&T broadband subscribers, including video, music, e-mail, and an enhanced version of the My Yahoo personalized content portal and RSS reader. The marriage between the two companies grew when AT&T became a reseller of Yahoo's paid search (Yahoo Search Marketing) through its directory sales channel, which sells print and YellowPages.com ads. An analogous relationship exists between Microsoft and Verizon in which

MSN provides content and services to the telecom's broadband subscribers. The companies then expanded this relationship to include distribution of Verizon SuperPages ads in MSN Live Local Search results.

The cohesiveness and evolving relationships between these companies indicate a likelihood that Yahoo and Microsoft will benefit from first consideration to bring content and communication platforms to federated bundled services. This will include proven and widely used on-line platforms such as e-mail, IM, and search. This will have the same advantages as those of Yahoo Go for continuity of service and content and targeted ad–serving capabilities. Cable companies could seek out these capabilities for their own bundled packages and will likely form partnerships with or acquire directory companies to bring local directional advertising into their on-demand offerings.

Because telecoms have relatively little experience in building and serving on-line communication platforms, they can benefit greatly from such partnerships or acquisitions. Portal platforms also come with the significant advantage of established user bases. Telecoms on their own would not be able to develop these technologies or comparable subscriber bases.

Such partnerships would also have marketing benefits. Picture, for example, the Yahoo Go mobile content product being called "AT&T Go." Consumers have an inherent trust in on-line brands for e-mail, IM, search, and other key platforms of federated bundled services, not to mention their inclination to continue using the same e-mail address, IM platform, and search engine. These are all highly habitual on-line activities, meaning portals have the greatest ability to accelerate the relatively slow adoption of mobile search applications by transitioning existing users.

AT&T has already formed a partnership with Yahoo Go by making the service available on some Cingular phones. The product, branded as AT&T Yahoo Go, is meant to enhance AT&T digital subscriber line (DSL) by giving customers mobile access to services they have on their PCs, including e-mail and IM. This is a big step toward triple play and the related content relationships that will unleash its true potential. In this case, Yahoo's incumbent position and millions of users give it an irrefutable value proposition in bringing the continuity of services across devices to AT&T. We expect that a potential MSN and Verizon union in the latter's bundled service offerings would offer the same advantages.

Smaller technology companies can also play a role in this ecosystem, as there are many pieces to the puzzle of networking, content aggregation, and content distribution over multiple devices. A team of AT&T executives recently took a day to listen to and vet a series of 18 presentations from start-ups, despite the telecom's tradition of working exclusively with large suppliers such as Lucent Technologies and Nortel Networks (which themselves traditionally subcontract with smaller companies). AT&T's goal in forming direct relationships with smaller start-ups is to bring innovative technologies to its main artery to facilitate the rollout of bundled services and the process of integrating many disparate technologies from SBC, AT&T, and the five other companies that have merged over the past eight years. Cable companies will likewise seek relationships with smaller start-ups that can bring innovation to the table.

"What start-ups do best is innovate," Menlo Ventures' Carolan said at the Dow Jones conference. "But to bring that innovation to market, you have to find a big partner who has a vested interest in its success. In a cable environment, they are battling with upcoming telco TV and they are battling with satellite now. They need these innovative features to try and differentiate themselves, and they are not going to start a new division to go do it themselves. I think in any of these ecosystems where there are a number of different parties along the value chain, any one of these start-ups will find themselves very well-aligned with some people and diametrically opposed to others. The trick is to find a good place to partner."

What about Google? "Net Neutrality" and Bundled Services

A major question at this stage of the discussion should be, where is Google in all this? The search giant may have made the mistake early on of not partnering with a major telecom, as Yahoo and Microsoft have, to establish an inside track for this next generation of services. Google has formed a partnership with Verizon SuperPages to have the Internet Yellow Pages resell its AdWords paid search inventory. However, Google's overall relationship with Verizon has not reached the qualitative level of MSN's, nor has it achieved ultimate relevance for bundled services (i.e., providing content and services for Verizon broadband customers).

Google, however, should not be discounted from potential telecom content partnerships, because of the value and strength of its technology, its marketing cachet, and its own growth in subscribers for on-line services. There is also the possibility of forming partnerships with cable companies in their bundled service offerings. Last, Google can still market its services to some degree where browser-based Web or mobile Web search is accessible.

Google's fate will also come down to a question of what will happen in net neutrality legislation in the near future. The net neutrality debate centers on whether telecoms can charge Internet companies and offer different tiers of service to distribute content over their wires. This compares with the current system, which presents a level playing field for any company that wishes to start a Web site and attract traffic. There are several net neutrality bills in different stages of legislative approval as a part of ongoing proposals to reform the Telecommunications Act of 1996. Resulting law would more decisively protect net neutrality. If telecoms get their way and are able to practice non-neutrality, Yahoo and MSN will indeed have a significant advantage in their existing relationships. If, however, net neutrality remains intact, on-line companies—most notably Google—will be able to compete across Web-enabled devices.

But even in this situation, a disadvantage exists in not being "baked into" the interfaces of IPTV service packages. The neutrality of the Web has allowed companies with the most compelling products such as Google to win market share in an openly competitive arena. IPTV conversely is a closed system in which all things must pass through the service provider's filter, causing relationships with the gatekeepers to be a source of considerable competitive advantage.

Similarly, real estate on a mobile phone interface is also valuable for portals. Just like on the Web, any user can access Google using a browser on a wireless application protocol (WAP) phone. However, the baked-in interface on a mobile device has better usability (for the same reason Microsoft Outlook offers a better experience on a PC than on Yahoo Mail, for instance). In the United States, carriers have a great deal of control over the hardware used on their networks, meaning a relationship with a carrier is a competitive advantage.

The closed-system approach for IPTV could incidentally end up harming service providers more than it helps them because it will prevent the advantages and the economies of the long tail to rule their service packages as they have done on the public Web. It could also cause consumers who want more choice to opt instead to watch video through the public Internet as they do today on sites such as YouTube.

This threat will grow as higher bandwidth makes bigger and better-quality streams available and as networking TV monitors with computers (to make on-line video viewable in the living room) moves beyond its current early adopter stage. Apple's recently announced iTV device, a wireless video-streaming set-top box the company plans to release in the first quarter of 2007, will accelerate this adoption curve. This will cause significant competition for IPTV service packages—a major reason telecoms are so opposed to net neutrality, as the data traveling for free over their pipes will threaten the IPTV service packages they are spending billions to build.

Net neutrality will be very important in shaping the landscape of content and advertising on the Web and its associated bundled services. But whether or not there is net neutrality, IPTV and mobile are more closed off to free competition when compared with the Web. That means the partnerships formed with bundled service providers in the coming months will greatly affect which portals have an inside track to becoming the content and advertising network for the next generation of on-line products.

Final Thoughts: Bundled Services + Targeted Ads = Local's Next Frontier

In the past, we have explored the targeted advertising capabilities of IPTV and the geotargeted local advertising possibilities of mobile search (see the June 12 ILM Advisory, "Targeting Users: Application Level Innovation in Mobile Local Search"). When combined with on-line local search, these will create exponentially greater opportunities to home in on consumers wherever they are using behavioral, contextual, and geographically relevant targeting.

This will include the ability to order products by Web, phone, or TV interface that appears in IPTV programming. The continuity of service across telephone, Internet, and television will also enable consumers to program and record video content via Internet or phone and use IM and videoconferencing through the TV interface.

"Convergence and integration are becoming increasingly important to consumers, and AT&T is building its strategy around three screens—the TV, the PC, and the wireless phone," according to AT&T. "This includes how customers can personalize television, how they may be able to access digital photos over the TV, and how they extend their TV experience to program a DVR remotely from another location and receive personalized stock, weather, and sports information."

On the local level, consumers will be able to order products, browse merchandise, and receive product information from stores in their area about items featured in TV programs. This opportunity will be amplified by the increased content choices of IPTV and the resulting new types and quantities of ad inventory, akin to the long-tail economics seen in on-line paid search. The on-demand and search capabilities that will be IPTV's flagship features will allow it to act more like classifieds, search, and Yellow Pages and thus offer content those directional media currently serve.

"My vision is that as IPTV takes off, we will serve up our Yellow Pages content database through it," said Dennis Payne, president and CEO of AT&T Directory Operations, at TKG's Directory Driven

Commerce event last month. "You will be able to search for pizza or home repair and get it right through the handset. We are working on this and have it built. We are beyond the prototype stage and the investment level is there."

These benefits will be most apparent when users can access their unique products and personalized features from any device. Most of the opportunity to serve the next generation of media consumers and target their unique patterns will therefore be realized by the companies that play a part in providing service bundles.

Internet service providers have erstwhile been overshadowed in content and advertising distribution by portal giants and nimble on-line companies that have proved better at developing technologies and user experiences. But looming net neutrality legislation combined with the relatively closed nature of the emerging services that will make up quad-play bundles will offer leverage to providers to play a more integral role in the content and advertising served on their networks.

As the portal giants that have ruled the Web will play a considerable role in serving this content, the way the quad-play battle shakes out could determine who will win the land grab for ad placement across all the communication devices we use.

Quadruple Play Can Make or Break MSOs: Steps to Prepare

Bill Bondy
Chief Technology Officer, Americas
Apertio

As telcos race to roll out Internet protocol television (IPTV) along with Internet access, voice over IP (VoIP), e-mail, messaging, and security services, cable operators cannot rest on their laurels by relying on their strongholds in the entertainment and broadband industries. Despite cable's solid brand recognition and established customer loyalty, telcos could gain considerable ground on cable turf by boasting "on-demand" TV capabilities and personalization of blended lifestyle services in their quadruple plays.

If IPTV subscriptions grow to 36.8 million by 2009, as predicted by Multimedia Research Group, this personalization will be a significant differentiator.

To stay ahead, multiple-system operators (MSOs) must recognize the many identities of a person as he or she transitions from personal, professional, and leisure profiles. A subscriber can be a wife, a mom, an office manager, a tennis player, an antiques collector, or a dancer at different times in the same day. The fact a subscriber could opt to change service settings according to the time of day, location, or situation could be leveraged to open the door to increased loyalty through improved service quality perception.

The problem is that embracing the customer and the seamless handoffs among TV, fixed telephony, broadband, and cellular networks will take substantial engineering feats. Of paramount importance will be the ability to instantly access information about bandwidth requirements, quality of service (QoS), permissions, pricing plans, credit balances, locations, and device types.

To achieve this, there needs to be a one-stop shop for data and a deep understanding of how dynamic services fit into rigid legacy networks with silo data storage structures.

While service management, control, and security can be greatly simplified with the unification of subscriber-specific data, the fact remains that multitudes of protocols and access methods go across many components (i.e., remote authentication dial-in user service [RADIUS], authentication, authorization, and accounting [AAA], session accounting, policy management, home subscriber server [HSS]). That makes consolidation a very daunting task.

With so many types of databases to manage—each with its own protocols and access methods—there is often a duplication rate of up to 35 percent. More often than not, manual processes and forklift migrations are the status quo for resynchronizing databases with networks in order to support and to keep up with increasingly rapid service changes.

Out with the Old, in with the New

The new-world view of data centralization is more dynamic, as it focuses on real-time capabilities and on-the-fly transactions. These capabilities require a move away from historical, report-oriented strategies that sat at the core of monstrous data warehousing initiatives and did not have rigorous latency and response time requirements. Monolithic libraries of information now have to give way to intelligent databases that "grip" data for deeper personalization of services and performance at increasingly higher levels.

To do so, cable companies have to break away from reliance on "transform layers" or "federation layers" that sit on top of multiple databases as an ad hoc "glue." While these layers help applications and clients to better understand the nature of queries, they will cease being real-time responsive and fall apart when dealing with 20, 30, or 50 databases. Because each data repository possesses its own access interfaces and protocols, the glue will no longer be enough when cross-database access in the network is required. Core network service and application performance lags are a major liability.

A centralized view will instead depend on the creation of one logical database to house all subscriber data with a discoverable, published common subscriber profile, as well as a single set of interfaces for managing that data (i.e., lightweight directory access protocol [LDAP], TelNet, simple network-management protocol [SNMP]). The single logical database will co-exist with data federation to allow a gradual, step-by-step migration of data on a silo-by-silo basis until the operator has consolidated all required subscriber data to the degree that is possible.

Get a Grip on User Data

Subscriber data is at the heart of control for the user experience and quality across networks. By consolidating customer data, MSOs enable provisioning and maintenance from one centralized location. A one-step process for adding all data for subscribers and services to a single database would give cable companies a huge opportunity to activate complex services within seconds of customer orders, rather than hours or even days.

Instant access to synchronized data will greatly improve the customer experience, as well as create tremendous operational expense (OPEX) and capital expense (CAPEX) savings. Potentially, miles of racks and servers could be eliminated if terabytes of data were moved to pizza-box-sized hardware rather than complicated storage-area networks (SANs) and larger servers. For example, to support tens of millions of users with less than a few milliseconds of latency on off-the shelf equipment would give operators a less than one-year payback if they go the data consolidation route for quad-play services.

For example, on the telco side, T-Mobile in Germany realized tremendous savings on floor space and equipment, as it was able to consolidate 88 HLR servers down to one 19-inch rack of equipment to accommodate 37 million subscribers. Reinforcing their service continuity, T-Mobile's subscriber data is real-time replicated over three cities—each site having its own 19-inch rack.

Key Elements to Data Consolidation

To realize similar CAPEX and OPEX benefits, there are certain components that are crucial to centralizing subscriber data among different network layers: a hierarchical extensible database; real-time performance; massive linear scalability; continuous availability; standard, open interfaces; and a common information model. To help prepare for the day when the IP multimedia subsystem (IMS) becomes a reality, leaving room for a software upgrade to a full-blown HSS will become important.

As cable operators integrate to PacketCable 2.0 environments, building and maintaining a subscriber-centric architecture will be key to services that require very fast, reliable, and resilient repositories that concurrently serve multiple applications. After all, latency is not tolerated in pre–IMS networks today, which could spell doom for quad plays that do not build on a consolidated subscriber-centric architecture.

A network directory server (NDS) is the first step in freeing and directing customer data from silos, as an NDS puts a directory in the heart of the network. With a centralized repository, service logic can be separated from subscriber data, enabling a cable operator to have VoIP and associated services working on wireless fidelity (Wi-Fi), because the subscriber data can be reused among various access networks (i.e., VoIP on cable, code division multiple access [CDMA], Global System for Mobile Communications [GSM]).

Additionally, the application independent and hierarchical nature of an NDS makes it extremely flexible and extensible, as well as suitable to host data for multiple applications and multiple access networks compared with embedded relational databases. A proper NDS directory structure is better suited to the disparate nature of the data prevalent in converged networks, which involve dynamic, real-time relationships. An NDS directory is object-oriented in nature with a data model that is published, enforced, and maintained by the directory itself.

For a network directory server to provide these capabilities in the core of the MSO network, it is critical that it be highly performant, massively scalable, and geographically resilient.

Typical disk-based databases and legacy directories do not offer the read/write speed operators need to consolidate data in a live core network. Average latencies of three milliseconds for a query and less than five milliseconds for an update are critical to maintain customer performance expectations. Update performance is critical, and using highly distributed memory resident directory databases can offer update (as well as query) transaction scalability at the point of access.

As critical as performance, a consolidated single logical database must always be available; downtime is loss of business. The network directory must provide continuous availability even in the event of multiple points of failure throughout the network, ideal for geographically dispersed networks and business continuity reassurance. NDS technology can be scaled massively, using data partitioning and distribution to host virtually unlimited quantities of data. Transactions and resilience are scaled by replicating data in real time over multiple local and geographically distributed servers.

To make this scalability cost-effective, the hardware must be compact, inexpensive, and non-proprietary and the NDS software must be able to scale linearly with the hardware. In fact, the hardware

necessary for high transaction rates with the aforementioned low latency is actually very small. A small network directory system can yield 10,000 transactions per second for a couple million subscriber data profiles on a handful of dual-core processor servers running Linux.

That is a big difference from relational systems, which rely on expensive and complex hardware to scale to high transaction rates and directory sizes. Relational systems often struggle to utilize more than a single server or operating system footprint to scale capacity, forcing much more expensive hardware into a network. That increases OPEX and CAPEX. Relational databases do have their place, as they are more the ideal for batch-mode, complex billing- and customer-relationship management (CRM)–type operations, but for voice services, short message service (SMS) and Internet services, distributed in-memory directories are more adept at handling the real-time nature of use when and where the data is needed.

Directories also help to simplify integration by supporting access through common IT technologies and protocols such as lightweight directory access protocol (LDAP), extensible markup language (XML)/services provisioning markup language (SPML), and simple object access protocol (SOAP). Using information technology (IT) technologies and protocols broadens the pool of qualified professionals who can support such as system. This translates into substantial cost savings, as operators can implement open interfaces in off-the-shelf hardware and operating systems. It is important to keep network components adaptable to a wide range of equipment to bring down support and maintenance costs.

Furthermore, to realize all the benefits of an NDS, it is critical that forethought be put into designing a common information model (CIM). This is the foundation for a useful, extensible data model that encourages data reuse while allowing applications to peacefully co-exist in a multi-application, single logical database environment. The CIM focuses on arranging subscriber, network, and application data in several categories: subscriber identities, common shared global data, application specific shared data, and private data.

Unfortunately, no standard model exists, as every operator has its own information model and its own methodology for migrating and consolidating applications. However, most MSOs can build a common data repository in their networks using an evolutionary approach. Starting with a single application that fulfills an emerging need of the MSO (e.g. presence, instant messaging [IM]), the CIM data model framework may be established. This provides the foundation upon which other application data may be integrated and built. From then on, new applications (e.g., Wi-Fi, AAA, policy management) can build on the already existing model. The key is to establish the proper foundation first and then add to it in an incremental fashion.

The CIM allows cable and telco operators to share data in a single logical database, as it houses reusable data that can be used for new applications and services. As new applications are added and existing ones evolve, data models are analyzed and changes are often required. Changes can be applied to existing application data models where data is part of a common model using virtualization techniques. So-called virtualization is the ability to provide application clients with different views of the common data based on the identity of the accessing agent. This allows the common data model to be filtered, reorganized, or enhanced to fit each application client's requirements while keeping the core data model intact and un-entangled with a specific application.

As data is "virtualized," objects can be viewed according to different characteristics. For example, attributes specific to a particular application or object distinguished names according to the accessing application or user. That means data is implemented once and managed as one instance, but it can be viewed as an object according to different characteristics over and over again.

As the CIM evolves, cable companies will need to find the synergies so that applications can share common data. Once you have shared objects, you continue to evolve the process of designing schema for applications and merging the schemas together into the common model.

The Next Step

As operators consolidate their subscriber data, the platform they choose must offer a seamless migration to support IMS data via an HSS. This prevents an operator from deploying yet another silo if or when the operator decides to deploy IMS. An HSS can also source its data from the NDS, storing its data as part of the CIM and thereby allowing IMS applications to source their data from the NDS as well as non–IMS applications. This has the potential to provide non–IMS and IMS applications a way to provide common data and services across different access planes. An HSS essentially sits on top of the NDS to offer a continued evolution to the process of consolidation. It does so as it enhances the CIM with an operator's IMS subscribers, the characteristics of their connected devices, and the preferences for those services.

Using an HSS in conjunction with an NDS as the master user database makes it the logical application and subscriber repository for the whole of the network. Each application can access user-specific data from a single source, so operators will not have to support specific requirements of each application's data store. This greatly simplifies provisioning and maintenance as well as data sharing. How to choose an HSS? Many options exist, but looking for one based on a highly distributed, replicated, subscriber-centric architecture will give the operator a platform that will evolve with their subscribers, applications, and networks.

Planning Consolidation while Standards Solidify

While a full PacketCable 2.0 implementation is still a somewhat distant goal for the cable industry, there are mixes of pre–IMS and PacketCable standards that operators can use as a guide to get consolidation projects under way. As IMS and HSS technologies evolve to help operators integrate across both mobility and landline platforms, some sort of hybrid will be necessary to start a path toward true fixed-mobile convergence (FMC).

A hybrid of PacketCable 2.0 and IMS standards can help to simplify application plays, control plays, and network access plays into one data layer that will ultimately unify all subscriber data across network data or service data.

For cable operators to guard their markets against hungry telcos that are charging toward IPTV, Internet service, VoIP, and other traditional cable services, they must start planning how to streamline their networks for faster services rollout. To achieve a real quad-play set of offerings, consolidation of subscriber data for unified views of customer profiles across multiple services is essential. Taking careful note of what technologies are available and how to undertake this transformational task is the first step toward a truly flexible network for MSOs.

Delivering IMS to the Home

The Role of CPE to Provide Full Service Convergence

Peter Galyas

Chief Technology Officer
Tilgin

Dante Iacovoni

Marketing Director
Tilgin

Executive Summary

The evolutionary vision of all–Internet protocol (IP) has its roots in using IP to transport legacy services. Later, the concept of the triple play was founded when voice, TV, and high-speed Internet services were introduced as a bundled offering. Adding wireless service provisions to the bundle created the quadruple play. One of the key tasks for service providers as they move in this direction is to capitalize on the full potential that IP offers. Providing legacy services over an IP infrastructure is only the beginning. The real challenge for operators is to utilize network convergence to deliver service convergence, and in the process transition their business models from that which is currently network-centric to powerful service-centric models instead. It is here the IP multimedia subsystem (IMS) offers so much promise, not least for the home.

Service providers around the world are investing in IMS infrastructures at an increasing pace. Convinced of the long-term value of this emerging standard, they are preparing to roll out new fixed and mobile IMS–based services. Meanwhile, consumers are buying more and more computers, telephones, TVs and other electronic appliances for their homes. If consumers could more intelligently network these appliances together and connect them to the service-rich IMS environment, the worlds of entertainment and communication would come together, and everyone—consumers and service providers alike—would benefit.

To effect this merger, a new type of customer-premises equipment (CPE) is required—CPE designed explicitly to bring IMS services into the home. Existing CPE was engineered for simpler tasks and are often based on proprietary technology. They lack the power, flexibility, and features needed to successfully integrate and evolve with IMS. A new generation of home gateways, set-top boxes (STBs), and other devices is needed to give home networks reliable, full-featured access to IMS services. Consumer CPE must evolve from simple access devices to key service enablers.

The functional requirements for IMS CPE are clear: They must guarantee the service quality of sensitive real-time communication. They must protect the security of both home and service provider networks. They must deliver converged voice, video, and data services to both legacy and IMS–ready devices. And they must allow all types of services to roam seamlessly between fixed and mobile devices. In short, IMS CPE must bridge the gap between communications and entertainment, introducing voice into entertainment and video into communications.

IMS CPE must also give network operators the tools they require to ensure reliable, cost-effective service delivery. They must include flexible deployment options and comprehensive remote management that extends operators' presence deep into subscribers' homes. They must allow seamless upgrades of premises devices, both to promote efficient upselling of new services and to enable end-to-end services that grow over time. Finally, IMS CPE must evolve as service provider networks evolve—from today's pre–IMS phase to tomorrow's full IMS implementations—without major disruptions or profit-eating upgrades.

IMS and Service Convergence at Home – an Example

Powerful new capabilities can be created by uniting home networks with IMS. Imagine, for example, a home network that includes a television, wired and wireless telephones, and an IP STB with built-in digital video recorder. An IMS–enabled home gateway links the network to an IMS service provider. To begin, David (the homeowner) comes home from work. His dual-mode cellular/wireless fidelity (Wi-Fi) handset automatically registers with a Wi-Fi access point built into the home gateway, establishing David's presence. Eager to watch the news, David logs onto the TV using a fingerprint scanner built into the TV remote. The STB updates the gateway to reflect David's presence at the television *(Figure 1)*.

While David is watching the news, a call comes in and the TV displays full contact information: caller ID, personal phone book entry, etc. It is Lisa, and she is requesting a video call. David has a few options. He could, for example, use the TV remote to reject the call or send it to voice mail. Instead, he takes the call on a nearby analog phone. The STB automatically pauses the live news broadcast.

While they are talking, David accepts Lisa's video call request by pressing the # key on his telephone. Lisa wants to share photos and video from her recent vacation, so she activates picture and video sharing from her home media server. Together, Lisa and David view the photos and video on their respective TVs *(Figure 2)*. When the call is done, David hangs up and resumes watching the news, which has been automatically time-shifted by the STB.

Perhaps the most salient feature of this example is the expanded role of the television. Besides allowing David to watch the nightly news, the TV and its remote are used to manage the voice call, support a video call, and share Lisa's visual material—a form of conferencing. More than broadcast television, video on demand (VoD), or even IPTV, David is enjoying true IMS TV.

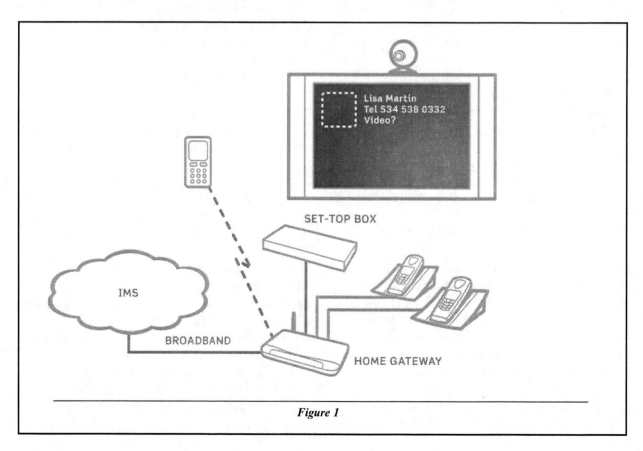

Figure 1

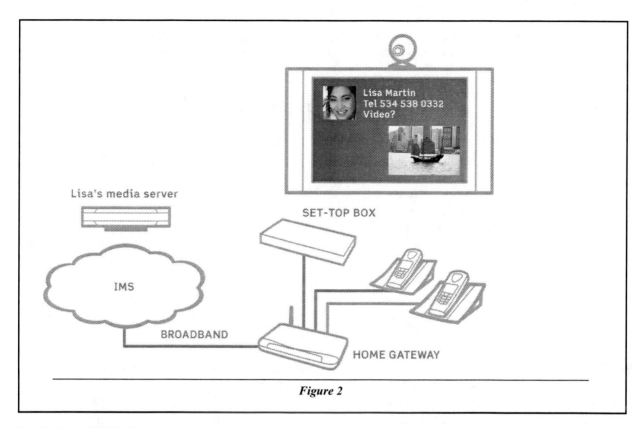

Figure 2

The Evolving IMS Market

The example above is compelling, but what will IMS really mean for service providers and their customers? Initially conceived to bring the flexibility of IP–based services to mobile networks, IMS is being embraced by service providers of all types as a standard architecture for next-generation network (NGN) services. The architecture comprises three layers. The network or access layer is responsible for access, transport, and interworking between carrier backbones. The control layer handles routing, policy enforcement, subscriber directories, etc. The application layer hosts media-rich services such as content sharing and presence-based communications as well as personalized and community services. Building on standard protocols such as session initiation protocol (SIP) and Diameter, this layered architecture reduces operating expenses and spurs application creativity.

Something for Everyone

IMS offers something for everyone. Instead of separate services delivered over separate access lines, IMS allows for a converged portfolio of IP–based information, communication, and entertainment services that boost service provider revenue and build customer loyalty. IMS affects all sorts of services that are currently carried over the public switched telephone network (PSTN) and second generation (2G)/third generation (3G) cellular networks, as well as cable and satellite. It facilitates the transformation of the public switched telephone network (PSTN), making voice over IP (VoIP) not only the dominant voice technology, but also an integral component of rich multimedia services.

IMS also influences the way consumer electronics are used at home and on the move. As IMS drives fixed-mobile convergence (FMC) far beyond voice, consumers will gain access to a feature-rich blend of voice, video, and data services from virtually any device: mobile handsets, fixed-line phones, personal computers (PCs), TVs, etc.

IMS helps service providers deliver an expanded portfolio of multimedia services and capture a greater share of consumer spending. Fixed-line providers, mobile network operators (MNOs), and mobile virtual network operators (MVNOs) are all looking to improve their service portfolios and competitive positions with IMS. IMS will be used for personalized, presence-based services that not only fight off fixed-mobile substitution, but also extend mobile offerings to include Wi-Fi voice at home and at hot spots.

Reinventing TV

Often associated with advanced voice services, IMS also brings special opportunities for managed TV services. IPTV is evolving, which affects services and puts new requirements on the IP infrastructure. Since all major service providers are investing in IMS infrastructure, it is natural for them to leverage their investment by adopting IMS for IPTV. The IMS architecture provides for quality of service (QoS), personalization, unified subscriber handling and billing, and integrated fixed-mobile convergence. By evolving from IPTV to IMS TV, providers will be able to offer subscribers more value than typically found in cable and satellite offerings.

IMS TV will enable operators to furnish much more than just video content. Because it uses SIP signaling, IMS TV supports combined services and interactivity for enhanced multimedia experiences beyond broadcast TV and VoD. IMS allows TV and communication services to share capabilities such as presence, instant messaging (IM) and profile management, creating a more personalized and community-oriented experience. Instead of just delivering one-way video, standards-based IMS TV will become a natural medium for both fixed and mobile entertainment and communication services.

Removing Limitations

Strictly speaking, however, many IMS capabilities are not new. There are solutions available today that merge data and video and let you surf the Web from your TV. But many of these solutions are proprietary, working with only one type of access network and one type of terminal device. The layered IMS architecture removes the following limitations:

- Services work consistently across diverse networks and devices—even across diverse providers' networks—extending their reach and multiplying their revenue-generating ability *(Figure 3)*.
- Service creation accelerates because application developers no longer need the cooperation of network equipment vendors.

- IMS standards extend the useful life of applications; network technology can evolve without requiring application rewrites.

As IMS Evolves

Of course, IMS will not happen all at once. Service providers need to confirm the new framework's integrity and value. They need to limit service disruption. And they need to manage costs. So service providers will introduce IMS in phases.

Today, most fixed-line providers are in a pre–IMS phase, moving rapidly from circuit-switched solutions to IP. IP–based softswitching, at least for Class-4 voice transport, is already nearly universal. Many fixed-line carriers are also offering VoIP services that have upgraded to SIP, the foundation signaling protocol for IMS.

Next comes early IMS, with its focus on shared control logic and IP–based service convergence. Core IMS components such as the home subscriber server (HSS) and call session control function (CSCF) are available today, and many carriers are conducting IMS trials. Still, full IMS, including key elements such as IP version 6 (IPv6) and IMS subscriber identity modules (ISIMs) for strong authentication, authorization, and accounting (AAA), are not ready yet for fixed-line access networks. So implementing IMS–based solutions is necessarily an evolving process.

Bringing IMS Home

Reflecting the smokestack architectures of legacy carrier networks, today's consumer communication services and consumer electronics are relegated to separate "islands" *(Figure 4)*. Routers link PCs to the Internet. STBs link TVs to digital TV services. Analog terminal adapters—standalone or as part of integrated access devices (IADs)—link legacy phones to VoIP services. Although there is some overlap, the networks, services, and devices are mostly separate. And while existing triple-play services can deliver telephony, TV, and Internet access over one cable, each service relies on a different terminal device. To achieve the IMS vision, multiservice capabilities need to be engineered into the total solution. To this end, a number of forums and alliances are working to specify standards, or at

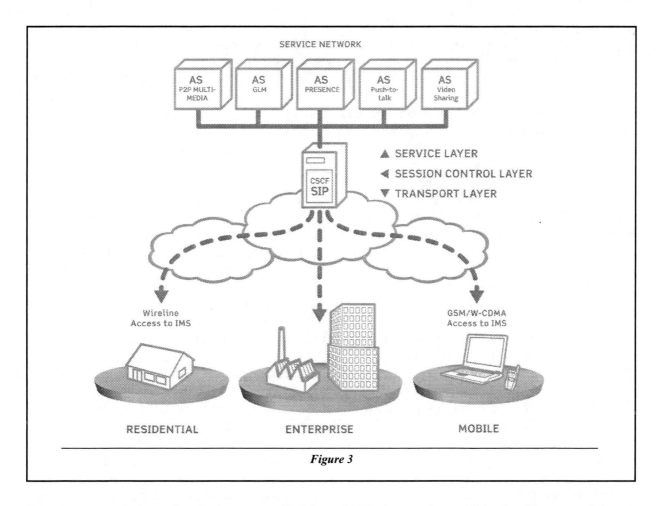

Figure 3

least recommendations, for devices controlled by the operator and the consumer.

Home Gateway Initiative
A coordinated effort by leading incumbent operators and CPE vendors, the Home Gateway Initiative (HGI) is creating a blueprint for state-of-the-art gateways that support IMS. To avoid vendor "lock-in," service providers and equipment manufacturers are cooperating to define the ideal home gateway and are making the results public. The first HGI document, released in July 2006, includes pre–IMS

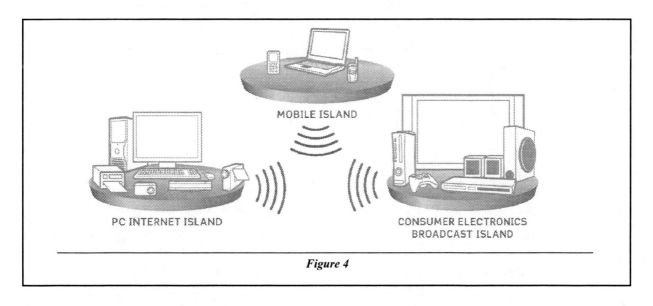

Figure 4

requirements. The next release, due in the fall of 2007, will map to full IMS.

DSL Forum

The DSL Forum is expanding its mission beyond digital subscriber line (DSL) to address interoperability and management across DSL, fiber, and alternative broadband technologies. Specifically, the BroadbandSuite Initiative is defining components that work together seamlessly to create the high-quality consumer experience that is vital for driving next-generation voice, video, data, and mobile services.

Digital Living Network Alliance

Through participation in the Digital Living Network Alliance (DLNA) and other initiatives, a number of companies are working to merge separate consumer electronics islands into a single cooperative system. In the DLNA vision, all sorts of home devices communicate with each other using IP over standard Ethernet, Wi-Fi, and Bluetooth networks. The universal plug-and-play (UPnP) device architecture automates device discovery and control.

Tying It All Together

Together these initiatives will create a standards-based home network environment that will not only spawn new applications, but also drive down cost. To complete the vision, the unified home network

must also be wedded to IMS, opening the door to a vast array of external services and resources *(Figure 5)*. And in order to "talk" to an IMS network, computers, TVs, phones, cameras, etc., must host IMS client software and eventually ISIMs.

An ISIM is an application running on a Universal Mobile Telecommunications System (UMTS) integrated circuit card (UICC) smart card and is the IP–based successor to the subscriber identity module (SIM) in today's cellular handsets. An ISIM includes the parameters and procedures for identifying and authenticating users and devices on the IMS network. The ISIM contains a private user identity (*username@operator.com*), one or more public user identities (*sip:user@operator.com* or *tel:+1-234-222-3434*) and a long-term secret used to authenticate and calculate cipher keys.

This specialized technology increases the complexity and price of individual consumer devices and threatens to lock consumers into products from specific vendors. Moreover, devices that cannot accommodate these enhancements would be denied IMS services altogether. Users interested in IMS services would be limited to certain telephones, media servers, video cameras, etc.

A New Generation of CPE

Designed for simpler tasks and often based on vendor-specific technology, existing consumer CPE

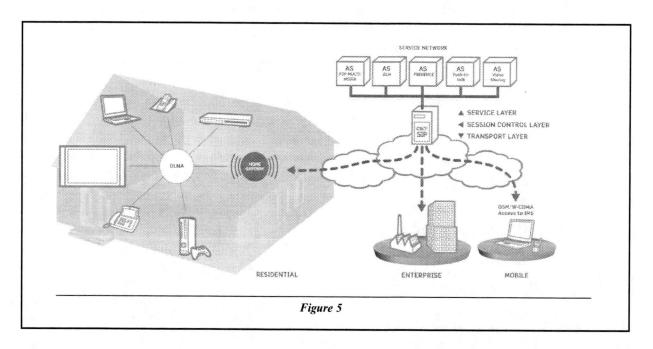

Figure 5

cannot be extended to support standards-based IMS services. Instead, a whole new class of IMS–ready home gateways, STBs, etc., is required to unite the home network with the public IMS infrastructure. These new devices would mediate signaling between home devices and the IMS network, act as IMS proxies on behalf of non–IMS legacy devices, and manage QoS in the complex multimedia local-area network (LAN)/wide-area network (WAN) environment. Because IP is fraught with security risks, the new devices would also host identity management, authentication, encryption, and related functions that protect service providers and their customers against hackers and other threats.

CPE designed specifically for IMS would extend the service provider network into the home. They would manage interactions between the home network and the service provider's back office, automating service deployment and software upgrades and enabling remote troubleshooting. The new CPE would be trusted on-premises entities for security procedures and network control.

Key Enablers

Purpose-built IMS CPE will be the key enabler for delivering IMS services to the home (*Figure 6*) for the following reasons:

- It will let subscribers access IMS services with their existing pre–IMS devices.
- It will reduce the cost of IMS services by relieving home devices of responsibilities such as ISIM security.
- It will accelerate service delivery, improve ease-of-use, and minimize service provider operating expenses (OPEX) by automating the provisioning and upgrade process.
- It will make the evolution to full IMS seamless and affordable.

Solution Requirements

To realize their full potential, IMS CPE must support service delivery capabilities for all manner of communication, including fixed and mobile telephony, IM, home media centers, IMS TV—

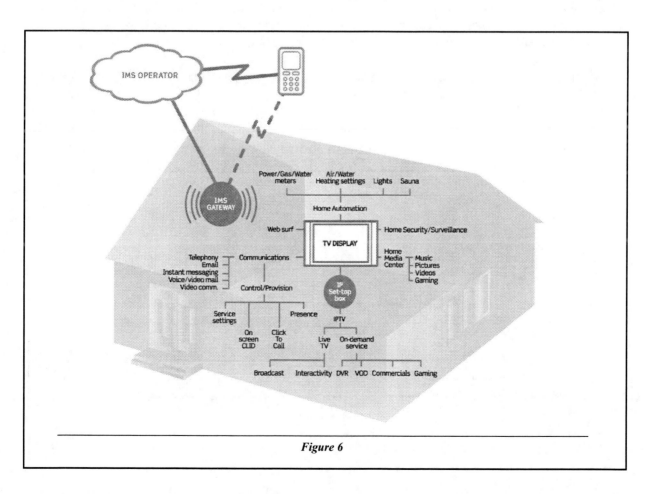

Figure 6

both broadcast and on-demand—and vertical applications such as home automation and security surveillance. The following solution requirements illustrate how IMS CPE would bring IMS services into the home securely and affordably and how it would evolve in parallel with service provider networks.

Versatile Configurations

IMS CPE must be highly modular and flexible in nature. Because service providers are implementing IMS in steps, the new devices must be designed to evolve in parallel. Depending on its role in the home network, each type of IMS CPE would assume one or more of the following functional responsibilities:

- IMS gateway client—Delivers base and supplementary PSTN–like services to legacy terminals (e.g., telephones, fax machines, modems), enabling cost reduction through replacement of primary PSTN phone lines.
- IMS multimedia client—Provides core IMS functionality to support fixed-line multimedia communication services that can be synchronized between fixed and mobile access networks.
- IMS proxy—Mediates SIP signaling between home devices and the IMS network and acts as a proxy on behalf of non–IMS devices. The IMS proxy registers with the IMS network and steers signaling and media traffic to and from

consumer devices as appropriate. It also manages QoS, AAA, and flexible mapping of public user identities to consumer devices.

- IMS TV—Supports the SIP signaling required to deliver and control broadcast and on-demand services such as pay TV and VoD. It also enables IMS–based unified billing and account management procedures with other home network elements.

Dynamic Self-Provisioning

Dynamic self-provisioning is a key feature of any IMS CPE implementation. Dynamic self-provisioning automates configuration and upgrade processes so that home users can add features and subscribe to new services unassisted (*Figure 7*). A user might start, for example, with simple broadband Internet access. Over time, the user could add VoIP service, IMS TV service, support for dual-mode phones, and so on. Each new service would be installed automatically, either by unlocking an existing software application or by downloading new software from the provider. The self-provisioning system should cooperate with the service provider's back-office applications and portals to ensure proper registration and billing.

Dynamic self-provisioning makes it easy to upsell new services. It improves customer satisfaction, which promotes new revenue and inhibits churn. In addition, dynamic self-provisioning reduces OPEX by eliminating truck rolls and keeping operator

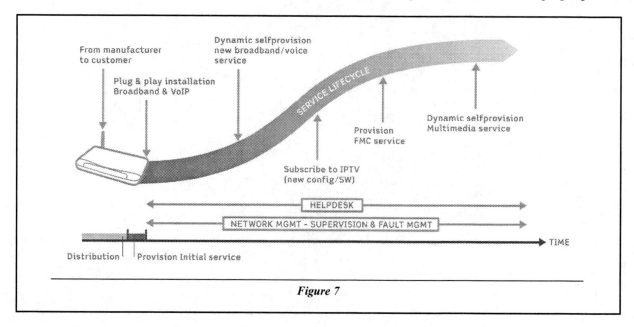

Figure 7

intervention to a minimum while still giving the service provider complete control.

Streamlined Service Creation

To promote new service creation, IMS CPE should support an application programming interface (API). A unified API would let service providers, system integrators, and software developers add value in the form of entertainment and communication software, digital rights management modules, etc.

A Solid Foundation

IMS CPE must be engineered specifically for deployment in the home. They must enable telcograde solutions that meet the ever-increasing need for converged voice and video communications. The following are some key characteristics that are required to support advanced IMS capabilities:

• Comprehensive remote management services that incorporate CPE and connected devices
• Granular QoS control that can align with individual applications in a multiservice offering

• Integrated digital video recording that includes digital rights management

Investment Protection

Because much of today's CPE is tied to specific services, operators have frequently based their purchasing decisions on a 12-month write-off or less. To support a new service, the provider not only purchased new CPE, but also accumulated costs for testing them, delivering them to customers, and supporting their installation. With so many changes on the horizon, this model no longer makes economic sense—if it ever did. Instead, IMS CPE must be designed to last and must evolve in parallel with the service provider's IMS implementation.

IMS CPE must work with today's pre–IMS networks, bringing VoIP, IMS TV, etc. to legacy devices. It must work with early IMS implementations to support basic signaling, routing, and security. And they must eventually leverage the features and benefits of full IMS, including ISIM–based security (*Figure 8*). Finally, IMS CPE must incorporate dynamic self-provisioning and automatic software upgrades that

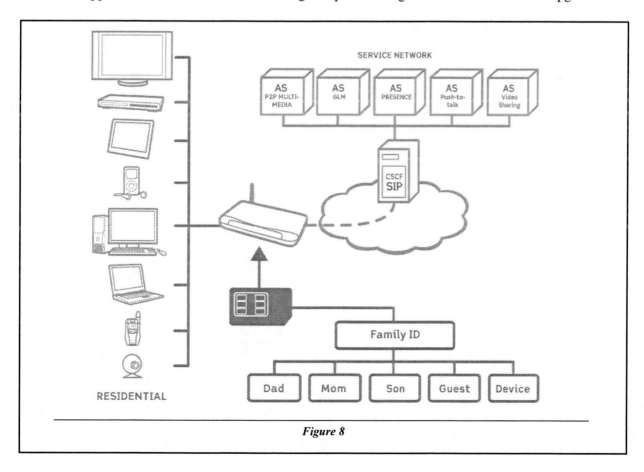

Figure 8

reduce the time and cost to distribute new software, eliminate disruption for subscribers, and increase the value of the service provider's CPE investment.

Summary

IMS promises a new era of anytime, anywhere multimedia communication. Efforts in HGI, DSL Forum, DLNA, and other groups envision a new world of integrated and personalized home entertainment and productivity. But this synergistic merger of information, communications, and entertainment will not be fully realized until home networks are wedded to the vast resources of the global IMS infrastructure. Existing CPE cannot support this union. Instead, a new type of IMS CPE is needed to broker the marriage—carrier-class CPE that extends the edge of the carrier network into the customer premises, creating a secure, cost-effective, manageable link that delivers powerful new applications and evolves in parallel with IMS.

Survival Skills in the Multi-Play Jungle

The European Perspective – New Ideas from the Old Continent

Marcelo Garcia
Convergence Expert

Welcome to Green Hell – Pay Media Operators Are Not Welcome

My grandparents were pioneers in middle of the Amazon rainforest during the rubber boom years early in the last century, before the synthetic equivalent started being mass produced. The nearest city was almost 1,000 miles downstream, where the Amazon River discharges 20 percent of the world's unfrozen fresh water supply into the Atlantic, a volume higher than the next eight largest rivers combined. There is more fresh water stored in the poles, but here you can actually go surfing where the river meets the ocean.

The jungle looks astonishing in those *National Geographic* pictures, but once you are down there reality bites—literally. Day and night threats of all shapes and sizes vie to shorten your lifespan, aided by stifling heat and near 100 percent humidity—it is no wonder that locals call the forest *inferno verde* ("green hell"). Because search-and-retrieval efforts are such a huge challenge, Brazilian legislation mandates that all aircraft authorized to fly above the jungle must be fitted with at least one spare engine.

If you happen to be a pay media operator or in case you consider getting into the great game of the new millennium, welcome to the multi-play jungle. Just to be on the safe side, may I ask how many engines your business is running on right now?

Like in the real jungle, many predators are lurking behind your back, figuring out how to profit from your weaknesses—but no matter how gruesome the struggle ahead, be certain that there are untold rich-es in stock for those who are prepared to face these challenges with the right infrastructure, knowledge base, and mindset.

However, this may require a drastic change in the way you operate your business (and perhaps also in the definition of which business you should be in), but keep in mind that even giants can manage the extraordinary—before the uplifting of the Andes in the Cenozoic, the Amazon River flowed in the opposite direction into the Pacific Ocean. I am sorry to disturb while you are working day and night, but if you pay enough attention you will notice that the Andes are rising right in front of your pay media business model at this very moment.

Not all is lost, brave explorer, because here we describe seven cases from just as many European countries, giving you their perspective on how to stay ahead of the curve and keep those nasty predators at bay. So get your khakis on to join our pioneers from Germany, France, the United Kingdom, Belgium, Switzerland, Denmark, and Iceland.

Framing the Problem – What You Know Today Will Not Suffice to Keep You Alive Tomorrow

Unless your home market happens to be North Korea, at some point in time increased competition from new entrants will put enough pressure on the pricing of your existing packages to push it below the break-even point. It is really not a matter of if but when it will actually happen. Bummer—got to get moving.

If you hire the best brains in the business to help you understand how to tackle this challenge, they are likely to come up with some of the following conclusions in those exquisitely bound final project reports that sell for thousands of dollars per kilo:

- After the broadband and voice over Internet protocol (VoIP) shockwaves, pay media is quickly becoming the focal point of the digital revolution, with new players rushing in from multiple segments (e.g., telcos, utilities, municipalities) all trying to have a slice of the viewer pie. Your pie.

- Audiences are becoming fragmented into "markets of one"—so long live the long tail of pulverized interests. Can you tell me which media portfolio would suit that entrepreneur with an M.B.A. who watches Bloomberg while planning her next rock-climbing trip to the hip-hop tune of the day? Now let us figure out what we can sell to her equally hyperactive husband and kids.

- Customers are naturally reluctant to embrace new services that are not already part of their lifestyles, although most are more than willing to do so after being convinced of their true value (e.g., TiVo, iPod). This is only relevant if we assume that your reserves in the bank will allow you to sit and wait.

- Premium content is king, and like most monarchs, it is expensive to acquire and manage. Hollywood studios probably will not even say good morning to you until they see a $100,000 down payment on the table. Prices for exclusive rights to major sports events make you laugh because crying does not help.

- If you are an European cable operator, on average you are a tenth of the size of your telco incumbent both in terms of revenues and subscribers (the ratio goes up to 1:23 in France). This is bad news in a game where scale truly matters and where telcos have huge war chests to defend their turf.

- If you are a digital subscriber line (DSL)– or fiber-to-the-x (FTTx)–based operator, that grin on your face is going to disappear when the cable guys start deploying data over cable service interface specifications (DOCSIS) 3.0, offering vast amounts of bandwidth for a fraction of the end prices sustainable through the aforementioned technologies at current equipment cost levels. Goodbye ROI.

- It turns nastier when wireless broadband gets into the rink—for instance, Verizon is spending $15 billion to bring fiber to 10 percent of all American households, while the whole country could be blanketed with worldwide interoperability for microwave access (WiMAX) for an estimated $3 billion in infrastructure and labor. Nothing in the game beats fiber for sheer speed, but this is not rock paper scissors—this is about making money in a sustainable way.

These are all wise, well-researched, and strongly supported arguments that probably keep you awake at night every now and then, making you wonder if you should work even harder or dust off your CV instead.

However, knowing the problem gives you only half of the solution, so the rest of this document will address several real-life initiatives from all over Europe that could help you understand how those issues are being addressed by thought leaders today and help you buttress your defenses before facing the battle ahead.

German Vorsprung durch Technik – Top Engineering Making Interactivity Accessible to All

Country demographics
- 82.4 million inhabitants
- 38 million TV households
- 69 percent cable penetration

Germans are quite conservative and tend to frown upon novelties even when they are obvious improvements to existing services, making their market a very tough nut to crack. However, past innovations with a clear customer benefit, convenient distribution networks, and a low price point have done extremely well over time (think "beer").

The Bundesland federation structure that has been used in the last 60 years has many positive aspects, but simplicity is certainly not one of them—to broadcast nationwide through ARD or ZDF (the equivalent of PBS), get ready for very long meetings with all 16 commissioners. It looks terribly tempting to go for the biggest fish in the European pond until you realize that you are actually going after a school of far smaller minnows.

So when in Germany, do as the Bavarians do. Munich is a hotbed of incredibly innovative technologies that never get publicized abroad because apparently no one bothers translating the top player's Web sites into other languages. Berlin has the glitzy government buildings, and Cologne has most of the media empires, but the land of the Oktoberfest is a powerful magnet to savvy high-tech entrepreneurs and qualified folks hungry for leading-edge action.

Take as an example this company, a serial start-up machine whose latest invention is an intelligent interactive remote control that works with virtually any TV set. It not only replaces your old remote, but also adds a mobile phone-sized screen and a few additional buttons for the interactive functions. An adapter that looks like a case for reading glasses plugs into your SCART connector (standard in all European TVs since 1995) as a pass-through monitor that interprets the signals inserted in the video stream at the broadcaster's facilities. Another box the size of a pack of cigarettes (connected serially) hangs happily from your phone line to report daily user activities back to your application servers—chatting with effectively any International Telecommunication Union (ITU)–compliant public switched telephone network (PSTN), integrated services digital network (ISDN), DSL, or VoIP line. Both adapters communicate with the remote control itself through radio waves, using an unlicensed frequency (around 431MHz in the European Union), hopping within the several channels available whenever noise from other unlicensed devices is detected (e.g., weather stations, remote garage door controllers).

Universal access allows interactive TV to finally reach the masses, especially those who are perfectly happy with their analog sets (insert "high-income older folks" here), that are cut off from the latest trendy developments. However, this is still only half of the story because the average "Hans Muster" ("John Doe") could not care less about novel technologies—all he wants is entertainment and information to match his lifestyle, and keeping him happy is a never-ending uphill struggle. So these sharp Bavarian gentlemen rolled up their lederhosen and came up with the following unique media cocktail to satisfy customers, advertisers, and broadcasters:

- Even if viewers have an electronic program guide (EPG), channeling interactivity to the remote eliminates the visual pollution caused by "red button" functionality, which hogs a substantial amount of TV screen real estate.
- Inserting the signals into the actual video stream enables perfectly timed and frame-synchronized messages to be sent to users—amusing, informing, or reinforcing an advertisement's message.
- Customers pay only around $50 for the device itself, a price similar to any fancy replacement remote. Premium events (always optional) cost a few cents, always equivalent to what is charged by short message service (SMS) and 1-900 alternatives. The return path uses a toll-free line, and an IP variant skipping the need for a phone line is in the pipeline.

So the good news is that not all is lost in the battle for "low-tech" customers when a device they have been very comfortable with for many years can be used to deliver effective interactive TV solutions.

DTT à la Française – From Patents to Profits

Country demographics
- 62.5 million inhabitants
- 22.9 million TV households
- 38 percent cable penetration

The French have a unique culture that prizes long-term views developed over never-ending meetings while paying an incredible attention to minutiae. The results come in the shape of the TGV—which has never had a fatal accident in more than 25 years zooming across the country at 300 kph—and some of the safest nuclear power plants in the world. Whether you like their style or not, you have to admit they must be doing something right.

The company described here applies this time-honored long-term thinking to the French DTT market. To achieve a breakthrough positioning, they have not only worked hard to conceive a state-of-the-art product line, but also kept their eyes peeled to what media legislators are churning out of those stern parliaments in Paris and Brussels.

A European Union directive from 2002 states that all digital TVs (DTVs) larger than 30 diagonal centimeters must be fitted with a digital SCART such as a digital video broadcast common interface (DVB–CI) module, i.e., a PC card–shaped unit. This legislation complements another directive from 1995 mandating the implementation of the original 21-pin SCART/Euroconnector interface in all European TVs. This 12-year-old law has effectively standardized E.U. retail video equipment the same way that RCA/cinch connectors facilitate connections elsewhere in the world, but with a superior output quality. Curiously, although the SCART interface was invented in France (it stands for Syndicat des Constructeurs d'Appareils Radiorécepteurs et Téléviseurs), it is called "Péritel" in its birthplace.

Another (very) recent legal development that reinforces their business model is the French "TV of the Future" initiative from March 2007, which will oblige manufacturers to sell only DTVs in the local market, going a step further into standardizing DVB–CIs at a national level and bypassing the need for DTT set-top box (STB) eyesores.

This company has wisely chosen to leverage patents acquired for their range of DVB–CI MPEG–4 decoders and will shortly launch a national DTT service working in close partnership with DTV manufacturers to benefit from their widespread retail distribution channels and customer goodwill. MPEG–4 decoder cards will be bundled with TV sets and include a free trial period, and reasonable fees will be charged to customers who decide to keep the equipment and profit from the premium channels to be available on the platform.

For reasons such as the lack of a true return path (unless coupled with DSL), DTT is bound to remain a niche player in most advanced countries, with a market share below 10 percent. What this company has decided to do was to target the crème de la crème of users with money to buy an expensive DTV set and a keen interest in cutting-edge technologies to justify that decision—probably the most attractive niche market of all in terms of revenue generation. To put a cherry on the crème, their hardware is very well protected by a set of global patents, raising entry barriers to virtually unreachable levels.

So the recipe for riches in French high-tech kitchens seems to be the following:

- Cook a very well-planned long-term strategy with strong fundamentals, based on favorable legislation changes, while applying the most advanced techniques that can be effectively protected by worldwide patents.
- Invite to the occasion the best manufacturers in the industry to leverage their brand recognition and extensive distribution channels while helping differentiate their product lines in the eyes of end users.
- Relax and enjoy sipping a nice glass of wine while your high-income, technology-friendly customers pour in—sustaining the low-risk, steady organic growth of your subscriber base.

Dessert, anyone?

British Pizza Boxes – Delivering the Right Ad to the Right Viewer at the Right Time

Country demographics
- 60 million inhabitants
- 25 million TV households
- 26 percent cable penetration

There are no Italians in this story—this is a very British company that has developed an addressable advertising IPTV solution so compact that it fits in a single rack unit (thus the "pizza box" moniker) while dishing out up to 200 individual video streams in a highly scalable stack configuration.

This is a pure IPTV play, but one worth a closer look because it changes quite a few game rules. For instance, satellite broadcast is a technology that boasts an incredible footprint (only Internet TV can do better) and has effectively 100 percent penetration in their territory, but it cannot dare segment markets without huge cost implications (i.e., transponder space). A few cable companies have been trying to segment their customer base geographically for many years with addressable advertising, but because the underlying technology has been very expensive and cumbersome to operate, it never really took off until our friendly IP packets came to the rescue.

We all love IP for one reason or another, but advertisers just adore the potential to send targeted messages to individual households and potentially even to individual customers if they choose to identify themselves while watching (e.g., by using a "personal" remote). Another great feature that has corralled advertising dollars into the major Internet search engines is the measurability at different stages of the ad consumption, which provides immediate feedback that in principle can be used to adjust the message being delivered not only in real time, but also in an automated fashion through the use of artificial intelligence (AI) algorithms.

The accuracy of the IPTV systems is truly remarkable—privacy concerns set aside (this is clearly a "permission marketing" scenario), once end customers have established their profile there is no need for a dog's owner to ever see a Whiskas advert again. That feline ad space will be allocated instead to another company—either a cross-market fast-moving consumer goods (FMCG) giant selling soap or a niche advertiser offering canine vitamin supplements that will make your rottweiler bite even harder. All this wizardry has been estimated to multiply your ad revenues by a factor of up to seven through the intelligent use of this extremely scarce and time-sensitive resource.

As you might have noticed, this addressability is decoupled from the geographic location of the viewer—a very dear Web 2.0 concept that allows even ubiquitous adverts to follow us wherever we are, sponsoring our precious content as always—but this time magically matching our lifestyles and minimizing customer backlash by showing them only relevant commercials. Even better, the whole plumbing takes place under your attentive monitoring room eyes because no new hardware or software is required at the customer's premises.

So if you are a telco getting into the pay media game where the local cable operator has been offering extremely low VoIP packages to convince customers to cut the copper to join their wonderful world of triple-play wonders, this is a way to help stem the inevitable churn through an enhanced customer value proposition while increasing your revenues by charging premium prices for premium ad space.

So now you know why those British gentlemen can keep their stiff upper lip sealed no matter what happens around them—they have done their homework before going out to play and came up with a platform that allows them to start a completely different ballgame. You would just wish that they invented a cunning way of explaining to mere mortals what the whole cricket fuss is about. Blimey!

Another Belgian Battle – A WiMAX Warrior in the Cable Kingdom

Country demographics
- 10.5 million inhabitants
- 4.5 million TV households
- 90 percent cable penetration

Belgium is famous for chocolates, cartoons, and cable. With 90 percent penetration, you would have a hard time finding a place that is not served by one of the several operators active in the country. People expect basic cable services at home the same way they assume that they can get electricity, water, gas, a few walls, and a ceiling as standard fittings. The telco incumbent is also very active in the multiplay game, and DTT covers any missing free-to-air media gap—pretty impressive stuff, which is all great news for end users spoiled for choice.

So why on earth would a well-established U.S. WiMAX operator join the fray? Shockingly, they have kicked off their European expansion with Belgium, Ireland, and Denmark and have passed more than 2.2 million homes so far. These are some of the most densely cabled countries in the world, and still this invader fuelled by very smart money has decided to take the plunge and invest in a very recent and relatively unproven technology. Are we missing something here?

Belgium is mainly flat, making it ideal for a widespread WiMAX deployment. For those of you not aware of this relatively recent member of the wireless broadband access family, WiMAX belongs to the Institute of Electrical and Electronics Engineers (IEEE) 802.x series, which includes wireless fidelity (Wi-Fi). In its current specification it can theoretically be operated in frequencies between 2 GHz and 66 GHz. Around the world, frequencies close to 2.5 GHz and to 3.5 GHz are preferred when license granting is involved, while unlicensed use of the spectrum tends to gravitate around 5.8 GHz.

The concept of broadband wireless access has actually been around for a long time and quite a few multichannel multipoint distribution system (MMDS) operators are actively providing services in those frequency ranges, usually standard-definition TV (SDTV) channels in areas where cable coverage is not practical—competing with whichever direct-to-home (DTH) service might be available in a specific region.

WiMAX is no MMDS because it is reaching a level of maturity similar to Wi-Fi, with very strong support from most major industry players driving the standardization of interfaces. This has never happened with MMDS or local multipoint distribution system (LMDS), keeping equipment prices at unsustainable levels for mass deployment. When the largest chip manufacturers in the world make a commitment to incorporate WiMAX into their chipsets, you are bound to see a dramatic drop in interface prices in the near future.

In many ways WiMAX is an inevitable long-range extension of Wi-Fi and to my fellow road warriors, there is no need to justify its many benefits. Although it complements nicely the higher bandwidth provided by Wi-Fi hot spots with a far wider range, it does have a negative impact on the Wi-Fi operators that today benefit from a cushy monopoly or oligopoly in luxury hotels and similar public areas with a high usage of laptops. All of a sudden a new entrant in the licensed or unlicensed WiMAX frequencies can offer substantial discounts to the businessman with a fat wallet staying at room 431 and all that without ever having to sign an agreement with the hotel chain. It may be bad news for some smaller players, but it is great news for consumers.

Napoleon owned most of Europe at some point in time and was feeling pretty confident until he met his nemesis in the flat fields of Belgium. Is a WiMAX Waterloo to be expected? It is highly unlikely, because the setup costs of the infrastructure are almost negligible when compared to wireline alternatives and also because the enhanced convenience factors mean subscriptions can be offered

at premium average revenue per user (ARPU) levels. Not to mention that in the unlikely event of customers politely ignoring your offer and sticking to a high-fiber diet, you can always move the transceivers to a new area with a higher expected return on investment (ROI)—not exactly a trivial proposition when a cable plant is involved.

"Swiss Made" Virtual Market Consolidation – Conquering a Highly Fragmented Cable Environment

Country demographics
- 7.5 million inhabitants
- 3.5 million TV households
- 90 percent cable penetration

Switzerland is a remarkable little country for many reasons, one of which is its very well-publicized stunning Alpine scenery, with around half of the European peaks above 4,000 meters and the vertiginous cliffs that come as part of the package. Another is the far less known fact that there are roughly 250 cable operators in a country with 7.5 million people—an average of 30,000 souls per operator, an astonishing number for a post-industrial economy.

This incredibly fragmented market was created by a combination of uniquely Swiss factors, including valleys surrounded by high peaks, which makes it uneconomical for external competition to face incumbents; a "small is beautiful" mentality originating from the cantonal political structure; a strong entrepreneurship culture; and lack of both population and government support for further consolidation. So here we are today with half of the market belonging to a major international group, while the other half is shared by a mix of utilities companies, municipalities, and family businesses sometimes operating cable platforms from wooden shacks!

Fragmentation has obvious disadvantages, especially in a converged world—negotiating quality content with major suppliers is unrealistic without a reasonable customer base, and aggregators can easily make these transactions overpriced for premium assets. Although many alliances and subcontracting deals take place to circumvent these limitations, the inherent difficulties of the process are unlikely to go away in the future.

"Grüezi, Broadband TV" company, has implemented a new service that effectively decouples content delivery from "Web 1.0" geographical limitations by streaming channels on demand even to customers with less than 1 Mbps of bandwidth sourced through their local Internet service provider (ISP). The free (ad-sponsored) service was launched in June 2006 for the FIFA Football World Cup and so far more than 400,000 customers have subscribed, although their access is currently restricted to computers leasing Swiss IP addresses due to licensing requirements. This innovative commercial P2P broadcaster has a clear expansion strategy and, after the Swiss success story, now offers similar services in the United Kingdom, Denmark, and Spain.

Although free for the moment, their approach has several advantages that could be leveraged in the future:

*   Customers would be more than happy to pay a modest fee even for traditional free-to-air channels just to have the convenience of watching them wherever they are in the country, especially in an out-of-home context (e.g., in a cafe through a Wi-Fi provider, at work during a lunch break).
*   They would be even more satisfied with the possibility of receiving premium encrypted channels that may not be available in their local multiple-system operator's (MSO's) channel lineup due to frequency plan constraints. It sounds bizarre, but in the past there were cases of minorities camping for several days in the lobby of a Swiss MSO's headquarters to be granted access to channels in their language.
*   In a broadband TV platform, the channel palette can be easily expanded according to customer demand, tapping the potential of a huge variety of "long tail" niches—no more indoor camping required.
*   Far more flexible pricing arrangements can be introduced on the fly—e.g., six-hour subscriptions to a premium sports channel on a "big game" Saturday evening or alternatives in case your team has already been disqualified from the tournament.

A lot has been written about broadband TV, but this is an excellent real-life example of how to bypass legacy roadblocks through the innovative use of relatively mature and increasingly reliable technologies such as H.264 streaming and P2P while squeezing profits from a mature market and enhancing end users' lifestyles in the process.

Danish Artificial Intelligence – Rocket Science from the Land of Lego

Country demographics
*   5.4 million inhabitants
*   2.3 million TV households
*   68 percent cable penetration

Denmark is a small country with no major industrial conglomerates worth that name (unlike Sweden) or major oil reserves (unlike Norway). Regardless of those apparent disadvantages, even after losing their grand empire centuries ago, they have managed to sustain high levels of income that grant them an enviable quality of life. Selling modular plastic toys helps them foot those social bills, but what is their real secret?

One of the reasons is trade—the Danish fleet incorporates the largest container ships ever conceived by mankind, plying the world oceans back and forth. But the trick we care about the most is their evident ability to work smarter, to be able to leave work early as most people in the country do. Incidentally, it is a lot easier to work smarter when you are the spin-off of one of the largest information technology (IT) companies in the world and your team includes a few Ph.D.s who know a thing or two about AI algorithms and how to apply them in a customer self-care context as is the case of the company described here.

Customer care—who cares? If you are a pay media operator you probably do that with all your heart, because you are spending up to 30 percent of your revenue just to keep customers mildly satisfied (or at least slightly below boiling point). This is one of the major roadblocks on the way to a true multi-play implementation because every new feature added to the platform could exponentially complicate your support requirements through a series of intertwined cause-and-consequence relationships. And with an average help desk call costing up to €6 in western Europe, you could probably do without the additional customer grievances.

The words "exponential" and "sustainable" seldom go together in the same sentence, so if you want to expand your palette of services profitably, you have to figure out first how to do that while increasing the underlying complexity in a linear equation. The following is our Danish rocket scientists' self-care solution, based on the Bayesian Networks theory:

- Most customer self-care systems in use today are plain-vanilla FAQs structured as a decision tree, an approach clearly not powerful enough to facilitate the resolution of very complex, multifaceted issues that very often require expensive customer-service representative (CSR) support or even more expensive truck rolls to be resolved. Such systems also require substantial and time-consuming modifications when a new service line is added to the platform.
- Self-learning AI systems such as those using Bayesian Networks theory assess statistical probabilities between independent events that lead to far better decisions in the cases where the customer is deeply confused and provides conflicting information with key facts missing (sound familiar?).
- What Mr. Bayes did to become famous was to define a network that can be used to implement algorithms where the "next best question" is continuously being recalculated as the customer moves on to the next stage, effectively emulating the questions that would have been asked by an experienced operator.
- As the overall user base provides additional independent events, the network becomes even more knowledgeable about a specific situation and this remains true even if the customer decides to skip some of the questions—something that is not advisable in a static decision tree scenario. In many ways this is similar to your global positioning system (GPS) receiver, which recalculates the best route when you get off-track using the knowledge built into the whole system.
- Such AI systems also bring on board several incremental advantages, because the most complex customer calls can be transferred directly

to experienced CSRs, the cost of training new CSRs is reduced substantially, etc.

In a multi-play world where complexity becomes the norm, you want a system that helps you stay in control of your business while minimizing the amount of human intervention to avoid outrageous implementation and maintenance costs. The country code for Denmark is +45 in case you feel like calling them right now.

Warm Icelandic Hospitality – Making Guests Feel at Home Wherever They Are

Country demographics
- 0.3 million inhabitants
- 0.2 million TV households
- 36 percent cable penetration

Iceland is a slab of rock in the middle of nowhere, touching the Arctic Circle in its northern confines. Heating is not a problem because the country sits on one of the fiercest tectonic regions in the world, making spotting new islands around the mainland a national pastime (not quite—the last one, Surtsey, popped up in 1963).

So the 300,000 hardy sailors of this solidified magma vessel have to squeeze the best deal out of nature's meager spoils. Right now one of those gifts from Gaia is the fact that Reykjavík is roughly equidistant to New York and Moscow, which is one of the reasons why Reagan and Gorbachev met there to bury their Cold War hatchets in 1986. A more relevant consequence from our multi-play perspective is that being between North America and Europe allows them to sieve the best ideas from both sides of the pond.

More good news—in 2005, Iceland topped the OECD table with the highest broadband penetration in the world and, to make things even better, literacy is virtually 100 percent. So far we have on our list a miserable climate, a strategic location, a highly educated population, and essentially ubiquitous access to the Internet. Hmmm—all of a sudden, writing leading-edge software starts to sound like a great idea.

The company described here has done just that and has created an IPTV middleware architecture from the ground up, using the latest recommendation from major Internet organizations—effectively coming up with a buzzword-compliant platform. Well done, but lots of smallish middleware vendors can make similar claims, so this is not the reason why they have been graced with a mention in this paper.

The real reason is the adaptation of their core software, specifically for the hospitality industry, which according to the World Trade Organization (WTO) currently stands ahead of transportation as the largest commercial services industry in the world in terms of trade revenues generated (financial services and software involve a negligible movement of goods and rank lower). International arrivals have grown at a compound annual growth rate (CAGR) of 6.5 percent for more than half a century—the 1950–2005 growth curve is remarkably linear in every region of the world—topping $685 billion total revenues in 2005.

Industry size clearly matters, but the most important point here is that while people are moving around more often, they still want to keep their lifestyles intact. An indication would be the vast numbers of DTH STBs registered in northern Europe that end up serving their noble duty in the Mediterranean coastline due to the generous footprint of major European Union satellites—a neat solution to keep score with what is happening back home, but plain illegal in most cases because of country-specific content licensing agreements.

Enter IPTV. With a marginal investment in infrastructure, all of a sudden hotels and resorts all over the world can make guests feel like staying longer by preserving their lifestyle in a transparent manner. Even better is the following:

- In a risk-sharing partnership scheme with content providers and infrastructure vendors, hotels would pay only for what is actually consumed, just as they do with mini-bar items. This mitigates risk and makes traditionally low-tech players more willing to implement cutting-edge platforms.
- Property owners can also leap over several generations from their clunky in-room entertainment systems (assuming they have one) to a tutti-frutti cornucopia of services including dig-

ital TV/radio, VoIP, TV browser, multimedia jukebox, and express video checkout.
- All of these services can be seamlessly integrated into their existing property management systems (PMS) and hosted in an application service provider (ASP) model to minimize internal training and capital expense (CAPEX) investment requirements.

A low-cost deployment in a shared risk model that brings concrete benefits to my guests? I will have one please.

Thanks for the Tales – Now What? Adapting These Novel Concepts to Your Own Reality

The seven cases described above focused on the real-life implementation of recently developed technologies in the European pay media context and were outlined using the following chapter structure:

- Introduction to the cultural context of the country where the technology has been implemented, which may or may not be relevant in your own environment. This should get you tinkering about ways of interpreting what can be done with your current platform from a variety of angles.
- Definition of the competitive landscape around the technology being described, not necessarily limited to the country in which it originates. This should provide you with additional insight on potential pitfalls in case you are interested in leveraging some of those ideas.
- The business justification that led these companies to develop or enhance inherently risky new technologies in a cutthroat industry. This should give you some cannon fodder to convince your board to take the plunge and push the envelope to make your balance sheet shine with new profit centers.

There has never been a harder time to forecast trends in a business traditionally associated to predictable revenue streams that mainstream financial institutions relish, but there has never been a more exciting time to be in the pay media business for true entrepreneurs that have the guts to face the challenges ahead.

We now live in a massively parallel world where it is essential to try as many approaches as possible at a sustainable cost level to allow you to identify the true long-haul winners in a playing field where the goal posts keep on moving.

As stated in a recent bestseller, markets that do not exist cannot be analyzed. Therefore, empirical experimentation is the surest way to define the best path to follow. Figuring out what you can safely ignore is just as important as knowing what your core business is.

The knowledge of well-established facts that has kept the status quo unchanged for such a long time now matters less than the ability to manipulate new concepts to let you acquire insight where others see only information. I hope that this article has given you plenty of food for thought, while at the same time making you hungry for more.

Welcome to green hell. Welcome home.

The Pain of Quadruple Play and How to Relieve It

A Shift in User Expectations

Natalie Giroux

Chief Scientist
Gridpoint Systems, Inc.

The advent of quadruple play intensifies a shift that has occurred in users' expectations. Users are now deriving their expectations from the nature of the applications instead of basing them on devices or types of networks they are using.

Not that long ago, mobile phone users were happy simply to be able to communicate wherever they wanted. They were satisfied, even with call drops or noisy communications. With the technology improving and increasing competition between the service providers, cell phone users now expect voice quality and reliability equivalent to those provided on the wired phone, to such extent that a good proportion of users now rely solely on mobile phones.

The same phenomenon is now happening with voice over IP (VoIP). Initially viewed as an affordable means to make long-distance calls, service providers are now compelled to include VoIP as a replacement solution to traditional phones for enterprises and consumers, who in turn expect the same performance, reliability, and quality as their traditional phone services.

Enterprises are more and more relying on videoconferencing to do business. Again, the expensive videoconferencing apparatus are being replaced by Internet protocol (IP)–based desktop conferencing or even mobile videoconferencing. Regardless of the device or technology, these users still expect a quality acceptable for them to communicate effectively.

The user expectations are strongly influenced by their experience. Another example, Web browsing, became a commodity in an era where bandwidth was highly available throughout the network. Even a cheap best-effort Internet access connection provided more than adequate response time. However, as service providers add revenue-generating quadruple-play services, it is only a matter of time before these best-effort users start experiencing degraded services, with the frustration and customer-service issues that service degradation entails. The only way to keep these customers happy will be to add tremendous amounts of bandwidth to the network, unfortunately without generating higher revenues.

The Demon of Complexity

In order to survive the intense competitive pressures and minimize churn, service providers feel compelled to offer bundled services. Furthermore, to make matters more complicated, users will soon expect to use in a ubiquitous way any of their applications on any of their devices or media.

Networks have evolved into separate islands (e.g., wired, wireless, IP, voice, TV, transport), often managed by independent business units, and the transformation required to combine these layers to allow for effective bundling is extremely complex.

The Ever-Increasing Need for Bandwidth

Early Web-browsing applications ran more than adequately with 1 Mbps of bandwidth.

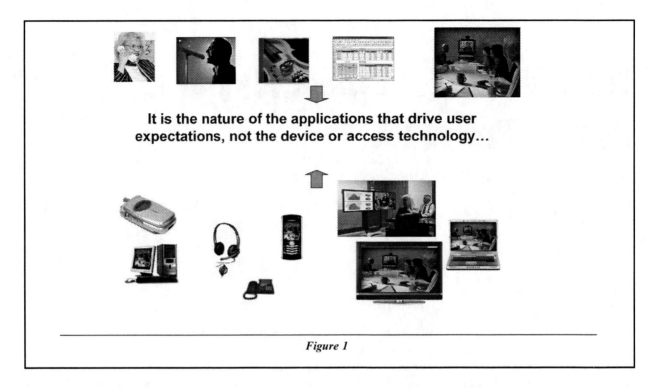

It is the nature of the applications that drive user expectations, not the device or access technology...

Figure 1

But quadruple play implies a significant increase of bandwidth for each customer. For example, the average demand per consumer is shown in *Table 1*.

This demand represents 26.2 Mbps of committed bandwidth plus 6 Mbps of excess bandwidth; this is more than 25 times the current allocation. Migration to 3G wireless comes with an additional sixfold increase in bandwidth.

Users who previously contracted their services from the providers with the appropriate expertise expect the same performance for a bundle, regardless of the original expertise of the selected service provider. For example, they contracted voice and teleconferencing services from a voice service provider, mobile services from a wireless service provider, data services from an ISP, and video services from an MSO or cable service provider. Some providers offered two or more of these services, but often

through separate business units. By developing expertise in one area, these providers raised the bar for expectations of service quality and performance.

To offer quadruple play, the traditional voice service providers have to build their infrastructure and become experts at supporting video and video on demand (including broadcast TV), even if that means a massive increase of bandwidth to each home or enterprise.

For multiple-system operators (MSOs) that already have the infrastructure to support video, they need to support the voice services with the same reliability and resiliency as the traditional voice providers.

Moving from triple play (voice, video, and data) to quadruple play, which includes wireless, adds another level of complexity and service issues, one main one being the physical limitation of the band-

2 VoIP lines	200 Kb
2 high-definition TV – VoD	2 x 8 Mbps
2 regular TV streams	2 x 4 Mbps
High-speed data	2 Mbps CIR + 6 Mbps EIR

Table 1

width. Wireless spectrum is restricted, and bandwidth achievable through radios is limited.

The small penetration and cost of reach with fiber will put constraints on bandwidth; competition is strong, and meeting end users' growing expectations is mandatory to stay in the game. The challenge lies in meeting these expectations in a profitable manner.

Solution

Network Convergence

In order to offer viable quadruple play, network complexity has to be addressed first. The complexity can be addressed only by removing layers of technologies in the networks and converging to one ubiquitous technology. There have been many attempts in the past to converge networks (e.g., frame relay, ATM), but the technologies were either missing some attributes (e.g., real time) or were not cost-effective. It is now evident that Ethernet is the technology of choice for the entire network technology convergence.

Ethernet will not only be pervasive in the access, but it will also soon be migrating in the metro and core, slowly replacing synchronous digital hierarchy (SDH)/synchronous optical network (SONET).

Network convergence will save service providers significant amounts of operating expenses (OPEX) and capital expenses (CAPEX) by eliminating overlap and wasted resources.

Ethernet is a flexible cost-effective technology already well entrenched in customer equipment (e.g., local-area networks [LANs]); it is also well designed to handle different types of traffic in real time and non–real time. Using a single technology across the network also greatly reduces the amount of bandwidth used for protocol conversion overhead.

In the past few years, there has been considerable effort made in the standards and forum groups (e.g., Metro Ethernet Forum) to improve Ethernet to add carrier–grade features to improve the manageability and reliability of the technology (operations, administration, and maintenance [OAM]) and make it ripe for prime-time convergence.

Planning and Provisioning – Bring Order out of Chaos

As service providers start to deploy triple-play and quadruple-play bundles, the next challenge is planning and provisioning the networks. Planning and provisioning are still tedious, manual, error-prone tasks, especially for large networks. Current

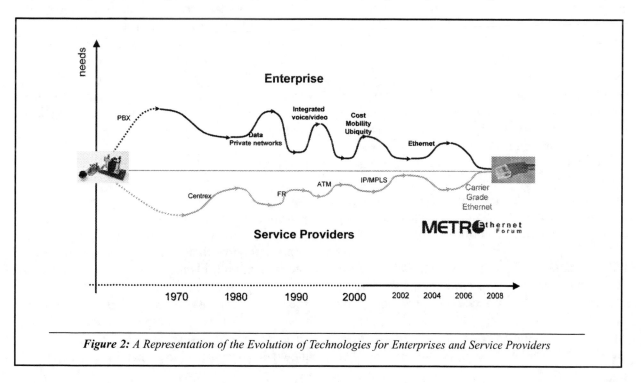

Figure 2: *A Representation of the Evolution of Technologies for Enterprises and Service Providers*

Ethernet routing algorithms use single constraints to find paths. However, service providers need to take into account multiple QoS constraints, along with custom policies for special circumstances, to make efficient use of their resources. Intelligent tools that can perform path selection, taking into account all the constraints and policies, are necessary to properly plan the network. As the network and the number of customers grow, manual input of scripts into each individual node becomes impracticable, and tools to automatically provision the paths become a necessity to eliminate costly errors and misconfigurations.

Protecting

As converged Ethernet networks expand to carry all the services for all customers, it becomes imperative to ensure full reliability and predictability. To make the converged network predictable and reliable, Ethernet needs to be implemented using a connection-oriented model such as pseudowires/multiprotocol label switching (MPLS) or upcoming standards such as provider backbone bridging with traffic engineering (PBB–TE) and transport MPLS (T–MPLS). Connection-oriented Ethernet (COE) removes the need for existing spanning trees and broadcast messages, thereby returning valuable bandwidth to support paying customers.

The COE model also allows for important enhanced OAM capabilities such as "bridge and roll," which permits an operator to test a path before sending traffic on it.

Quadruple-play bundles include various types of services (real time and non–real time) and, as such, the users, initially enterprises, will increasingly demand contracted service-level agreements (SLAs) with guaranteed QoS objectives. The service provider needs to set up protection paths for each service covered by the SLA to avoid downtime during failure and the penalties associated with failing to meet the SLA conditions.

With COE, failure protection with minimal service impairment becomes possible because the backup paths are pre-computed and pre-provisioned for each service. Furthermore, it allows service providers to implement network-aware traffic management that can proactively switch paths when

SLA violations are impending. In addition to hard-failure conditions, SLA violations can happen with adaptive wireless links or when the network is overbooked too aggressively. Such a feature as COE can truly protect the network and the services against any type of impairment.

Profiting

Once the network is properly planned and the services are provisioned and fully protected, the next step is to make the service offering profitable. This is done by optimizing the use of the bandwidth throughout the network by balancing the load and maximizing the use of the bandwidth for paying customers, as opposed to carrying retransmitted packets, as is the case in some networks with poor performance characteristics. As mentioned before, quad-play bundles need an increase of bandwidth of about 25 times the current allocation. Service providers have to add a significant amount of bandwidth throughout the network to address the need. To meet the users' performance expectations, overprovisioning (allocating more bandwidth than is sold) is currently necessary to avoid performance degradation due to congestion. But as customers demand lower cost for their services, this model quickly becomes unsustainable, and intelligent bandwidth allocation and traffic management become necessities. With COE and network-aware traffic management, it is possible to minimize the bandwidth allocated to each service and thus maximize the number of customers using existing bandwidth before spending more capital to add resources. These same techniques work very well with the new variable-rate wireless radios as they balance the load of the traffic on the network dynamically.

Summary

Quadruple play will be enabled by the emergence of connection-oriented Ethernet, combined with intelligent service management for planning and provisioning and network-aware traffic management for protection and profitability.

Expectations were at one time inflated by the overabundance of bandwidth in the network. The cost-to-performance ratio did not reflect the amount of bandwidth consumed. Now that the expectations

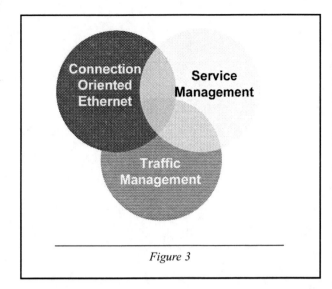

Figure 3

are established, service providers need a bandwidth-efficient solution to continue to meet these expectations while offering bandwidth-hungry applications. Simply adding bandwidth when applications require more and over-provisioning and over-protecting to ensure customers expectations are maintained is not a sustainable model. In order to be profitable, quadruple-play offerings need to be based on intelligent network planning, automated path provisioning, and failure and SLA protection, combined with efficient use of existing bandwidth. With such architecture, new revenues can be achieved by creating new applications, opening new markets, and reaching new customers. Significant cost savings can be achieved by more efficiently using capital investments and human resources. Costly bandwidth increases can be delayed until justified by revenues.

The intersection of carrier-oriented Ethernet, service management, and traffic management offers the ideal solution to improve user experience and guarantee customer retention.

Will Quad Play Be a Home Run or a Strikeout for CSPs?

The Answer May Lie in Your Customer Service

Jeff Gordon
Senior Vice President, Innovation Center
Convergys

The appeal of triple play or quadruple play to communications service providers (CSPs) is clear: Providing more services translates into bigger market share, more revenue, and higher customer lifetime value. But fail to provide adequate customer support for these services, and you may lose out on these benefits. Millions invested in networks, infrastructure, and marketing may ultimately be wasted if your customers cancel or fail to use services due to frustration or lack of understanding. This is just one reason customer service is increasingly becoming a strategic differentiator for CSPs.

Service bundling may ultimately benefit the customer and the CSP, but it poses an unprecedented challenge to customer service. This challenge comes at a troubling time for the customer care industry; statistics show that customer satisfaction levels across all industries have flattened or declined over the past few years even as spending on customer care continues to surge.

Merely adding staff or striving to improve existing metrics just will not cut it anymore. We need a new paradigm for customer care. In this new paradigm, customers will see the CSP as an active partner in solving problems and making life easier. To fulfill this goal, companies will use new channels and new technologies such as Internet protocol [IP]) to understand and interact with customers in ways never before possible.

The Missing Link in Convergence

CSPs are aiming at convergence, but the customer care paradigm must converge as well. At the moment, we are a long way away from that goal.

Multi-company partnerships are needed to build and maintain the devices, applications, and networks that make converged devices function optimally. With so many players, customers may not always know whom to call when there is a problem. And when customers do call, they are too often bounced like a pinball from agent to agent, or from company to company. At each stop they must wait in hold queues and retell their story. Then they must wait some more as agents scroll through pages of data in search of information.

For customers increasingly accustomed to solutions on demand, it can be a maddening experience. James Canton, the futurist who addressed our 2007 Convergys Executive Forum in January, predicts that "the increasing complexity and sophistication of products and services will increasingly tax the competencies of customer-care professionals, and will continue to frustrate consumers if they're not getting their needs met."

Changing the paradigm of customer care requires "converging" on the customer's needs, that is, reorienting processes and attitudes toward a customer-centered view. I believe the customer care center of the future will have four key elements:

proactive care, automation effectiveness, agent efficiency, and customer value optimization. This paper will explore what each of these will mean in practice.

Proactive Care

It is said that we live in the information age, yet contact centers often seem information-starved. Businesses gather terabytes of data on their customers' context, transaction history, and inclinations. Their staffs are experts on the benefits and flaws of their own products and services. Yet this knowledge is seldom used to improve the customer experience in real time.

Instead, businesses wait until a problem prompts frustrated customers to reach out for help. If customers do not feel their personal needs are being met via automation, they will demand attention from a live agent. Even if the problem is resolved, the customers' time has been wasted and the company has fielded an expensive call.

In the future, customer-care operations will use the information at their disposal to proactively reduce both costs and consumer aggravation. Businesses have more ways than ever to communicate with customers. So why not use them?

Imagine, for example, that a longtime customer of yours is being charged her first late payment fee. You know from experience that such customers are likely to call in, pay up, and request a waiver for the late fee from a live agent. So why wait for the call? In the future, this customer will receive an e-mail, text message, or automated call—before receiving the bill—notifying her that if she pays promptly through the company Web site, no late fee will be charged. This scenario saves the customer from frustration and the company from an expensive call.

In the future, contact centers will anticipate customers' needs and problems, improving satisfaction while reducing costs. Which customers need to be notified of a service outage, a product update, or a troubleshooting tip? A proactive contact center will know the answer and take action. And with the increasing prevalence of IP–addressable appli-

ances, it will become easier to locate your customers and address their needs in real time.

Automation Effectiveness

The goal of automation is to handle mundane tasks so that live agents are freed up for higher-value tasks such as selling while maintaining or even increasing customer satisfaction. This is another realm where making full use of the customer information lodged in your data warehouses is useful.

In the future, automated menus will be highly personalized. Say you are the customer of a bank. After verifying your identity, the voice response unit will greet you by name: "Hello, Mr. Smith, are you calling about your CD, which is set to expire in three days, or about your bank account, which has a low balance that might soon incur a fee?"

In the background, an automated program will monitor the quality of this dialogue. If there is a hitch—say, the computer fails to understand a customer, or the customer's voice rises in frustration—the program alerts an agent. The agent can speak to the customer directly or quickly solve the problem from behind the scenes.

Agent Efficiency

In the past, companies gathered data chiefly to serve their internal purposes. Information systems were built in functional silos to serve one part of the business. As a result, most customer service representatives sit before desktops loaded with a large number of applications and waste a lot of time fishing through them to find useful nuggets of information.

In the future, information will be organized and presented in ways that are optimized for customers and the front-line employees who serve them. Agents will have access to a comprehensive view of customer interactions over time and across multiple channels. This holistic view will allow agents to anticipate and resolve issues that might otherwise have led to future calls for help or to better spot opportunities for upsell, cross-sell, or retention efforts.

What's more, the "pinball game" will soon be over, as customers are provided with a single point of contact. By leveraging the next generation of voice over IP (VoIP) follow-on standards such as session initiation protocol (SIP) and session containers, a customer contact will be initiated one time and, as collaboration occurs among generalists and specialists, they will each "join" in the customer's contact session, where they see the customer's information and actions taken by other staff members during the session. Agents will also "blend" multiple channels in a single session. For example, an agent could "meet" a customer on the phone, then take over the customer's PC and guide the customer to a solution through co-browsing.

This concept can be extended to the full customer service value chain. If your company has partnered to deliver a service, you will want your partner's agents to appear to be yours. With collaborative care, your agent will be able to tell the customer, "Let's consult with the expert on that," rather than, "Here's an 800 number to call."

Optimizing Lifetime Customer Value

CSPs will fail to maximize the returns from these improvements to customer care unless they can tailor their sell, service, and save efforts to account for each customer's assessed, anticipated lifetime value (as well as to customer, product, and support life cycles). This is the best way to ensure that upsells, cross-sells, and the like occur in a timely and relevant fashion.

Optimizing lifetime customer value requires a systematic approach. That is, it is not enough merely to assess a customer's value or even to have business policies in place that dictate appropriate service levels. You also need an infrastructure that will allow you to make sure that those policies will be adhered to consistently not only through automation, but also through the agent's applications.

In conclusion, let us remind ourselves of why customer service is important. Early adopters or casual users of free services may be willing to tolerate poor service. But for CSPs aiming at the mass market, profitability will depend on building long-term relationships with paying customers. And even if adding services gives you an edge in the short run, in the long run, many services will become commoditized. For all these reasons, customer care will make a big difference in helping you stay ahead of the competition.

Provided to the International Engineering Consortium courtesy of Pipeline Publishing, LLC. All rights reserved. "Will Quad Play Be a Home Run or a Strikeout for CSPs?" originally published in Pipeline Magazine Volume 3, Issue 10, March 2007, can be viewed at www.pipelinepub.com.

Developing a Converged Delivery Platform for Quad-Play Services

Ron Iannetta

Senior Manager, Communications and Media
BearingPoint

Introduction

Today's telecommunications service providers are formulating plans to use their next-generation Internet protocol (IP) networks to deliver quad-play services. They realize that the combination of wireline voice, Internet access, IP television (IPTV), and wireless services in bundled offerings will create more customer loyalty and higher revenue. However, they are struggling to roll out these new services to customers over their new IP networks using the same basic systems that supported them in the past. Although next-generation networks (NGNs) are being deployed, next-generation services are lagging. These include voice communications that incorporate audio and video content as well as capabilities traditionally thought of as Web services, including presence, preferences, location, buddy lists, and address books. The challenge facing service providers is how to deliver these new services seamlessly and transparently to end users. What kind of underlying cohesive system architecture do they need to support new quad-play services over an NGN, and how should the functional systems interact with one another?

Service providers are looking at different solutions to address these questions. Wireless carriers are adopting a specification called IP multimedia subsystem (IMS) that they believe will provide the core infrastructure they need to deliver these new services. Web service providers are looking at service delivery platforms (SDPs) to provide a similar infrastructure. So far, however, neither domain has been integrated to the point where seamless converged quad services are possible on a single delivery platform.

To make matters more challenging, operational support systems (OSSs) that support most application services today over the same network tend to either be independent of one another or operate in a silo. This means that each application is a stand-alone service with its own features, user interface, and access security and incorporates features that may vary from application to application. The only way to deliver seamless converged services is to allow the underlying applications supplying these stand-alone services to become increasingly more integrated with one another.

What Are Converged Services in a Quad-Play Environment?

From a service provider's point of view, a service can exist at any layer of a telecom network. Historically, these services existed on the lower layers of the network—network layers 1 to 3 of the open system interconnection (OSI) model. Application services, on the other hand, operate on the application layers—specifically layers 4 to 7—that sit above the network layers. These services include e-mail, instant messaging (IM), voice over IP (VoIP), wireless, Web access, or IPTV. All of these and more are elements of a quad play bundle. For clarification, this paper may refer to these application services simply as services (or just applications) with the understanding that they are services provided at the application layers that use the underlying systems and network.

There are three classes of converged services that would exist as part of a quad-play bundle: communication services; entertainment services; and data communication services.

Communication services include wired or wireless voice telephony, voice mail, e-mail, and IM. As VoIP continues to make inroads into traditional wireline voice services and future wireless services, it will become more of a data service over session initiation protocol (SIP), incorporating data features such as presence and eventually offering capabilities that will transform it into fully integrated voice/data/video communication. Similarly, voice mail, e-mail, and IM will likely merge into a single unified messaging service that supports multimedia content.

Entertainment services, by their very nature, deliver feature-rich media content to a subscriber. Today, those include pay-per-view (PPV), video on demand (VoD), music or video downloading, streaming video, and on-line gaming. Tomorrow, these services will include IPTV, music on demand, and multiplayer gaming. Entertainment services can be on-demand as streaming applications controlled on the fly or downloaded and stored on either a network or local personal video recorder (PVR) for later access.

Data communication services do not fit into either of the previous two service classes. Examples include browsing, searching, information retrieval, software distribution, network-based application software, and the entire multitude of e-commerce applications.

How, then, can these services converge? Consider a user talking on the phone to someone. Both parties want to add another person to the call, but they do not know the number. They access their on-line address book, look up the number, and then select click-to-dial to automatically call the person and add him or her to the call. In other words, the first user in the call session seamlessly invokes a Web service (address book) and then has the Web service invoke a communication service to initiate the

call to the third party and conference that person into the existing call. This simple example is actually somewhat complex to complete successfully and requires integrated session management capabilities across both the voice and Web services support systems.

Another example involves someone who wants to know what restaurants are within a 10-block radius of his present location. He invokes a finder service on his cell phone, selects restaurants, possibly even selects type of food, keys in his search distance, and receives a list of possibilities on his phone. A restaurant is selected from the list and the system initiates a call to that restaurant to place a reservation. Again, this scenario encompasses both Web services (e.g., directory services, location, security) and calling services (e.g., Web service invocation, call initiation and setup).

These scenarios are examples of converged services—services that use traditional Web services and call services together in the same user session, blended seamlessly into a single experience for the user. However, current underlying support systems are struggling to deploy such converged services.

The Technology Effect of Today's Trends

Today, how and what to provide in the way of new services bundled together for a quad-play over a next-generation IP network are influenced by technology and marketplace trends. Some more than others will exert strong influence over how the underlying systems need to be designed to support converged services. These trends include the following:

As network architectures evolve, the services running over those networks are becoming more separated from the networks themselves. Historically, the services provided have been incorporated into the intelligent network equipment or closely tied adjunct software systems. But the trend is for these services to be provided by application platforms that are network-agnostic and independent of the underlying networks. This would allow a service to ride over a wireline network for one session but use a wireless network for the next. Or that same service could initiate on one network and complete on another in the same session. Wireline-wireless voice service is an example of the latter.

Services today are constrained by the lack of bandwidth in access networks that exist between the network backbones and the end devices delivering those services. Such performance constraints are changing rapidly as service providers continue to upgrade and evolve their access networks to meet new standards. For instance, in the wireline arena, providers are investing heavily to bring fiber closer to the endpoint of service (e.g., Verizon FiOS, AT&T Lightspeed). In the wireless world, service providers are adding bandwidth capabilities to existing sites and adding new ones as quickly as investment strategies and regulations allow. This will enable new services that require large-access bandwidth, including IPTV and streaming video, as well as heavy graphics applications such as gaming.

Consumers are becoming more mobile and want their services to follow them instead of being fixed to certain devices and locations. This means that service providers will need to enable their services to operate across different networks and devices. These services will need to interact seamlessly no matter on what device or where they are invoked. This becomes a real challenge as the systems supplying those services must be aware of the capabilities of each device type in order to present a service in a useable way.

Voice services used to be totally separated from data-oriented services, and the two would use different technology for delivery, albeit most often over the same underlying wireline or wireless network infrastructure. But voice services today are moving to IP–based infrastructures such as VoIP and will become just another form of prioritized data to be delivered over the same network. In fact, full wireless-wireline convergence will become a reality through the common use of IP in both wireless and wireline networks. This trend will allow voice services to be blended with other data services into a seamless environment.

New technology lowers the cost of entry for new players and creates a more competitive marketplace for service providers. They will need to find more effective ways to compete, including lowering costs

for new service development and shortening time-to-market windows. Moreover, not all new services will be successful in the market, and service providers will need to use rapid deployment platforms to test their services on a trial basis before committing to a wider rollout.

Service Application Architectures

As previously mentioned, services and their underlying applications will continue to be more and more separated from their delivery networks, even as they expand in scope and variety. Basically, two significantly different architectural models can be used to support service application development and deployment—silo models and layered models. The existing and widely prevalent silo model, depicted in *Figure 1*, can continue to be extended. Using this model, each application is self-contained, comprising all the capabilities that it requires within itself. It does not rely on external services or applications to deliver its services. But some real shortcomings exist with this model, despite the fact that it has been used for years. These shortcomings include the following:

- A set of capabilities over many applications are developed over and over again, resulting in unnecessary duplication of effort, time-to-market delays, and a waste of resources. This issue is probably the most compelling and the one most likely to drive change, regardless of how elegant another solution might be.
- The end-user experience differs from one service to the next, depending on how duplicated capabilities were built in each application.
- Seamless interaction across services is extremely difficult to implement.
- Service management happens in each application, so there is no single manager across all services that can integrate them.
- Because each application is different and self-contained, third-party applications would need to build separate interfaces to each one.

So how do we overcome these problems? The alternative to the siloed architecture model is a layered model, depicted in *Figure 2*. In this model, a set of capabilities used by several applications is implemented as a common services layer. Each applica-

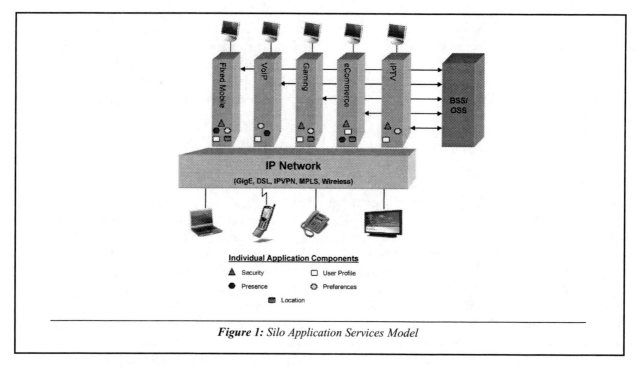

Figure 1: Silo Application Services Model

tion invokes these common services as required to deliver its own specific application service. Applications such as unified authentication and authorization, presence, availability, location, profile management, directory services, subscription management, session management, usage management, and interfaces to OSSs such as ordering and billing are a few of the services that can be built into this common services layer. In other words, this model includes common services required by each application, regardless of the specific service it supports. As depicted in *Figure 2*, this common services layer would interface with all of the applications that use it and the underlying network. It would also support a uniform customer interface as well as Web services interfaces to key OSSs.

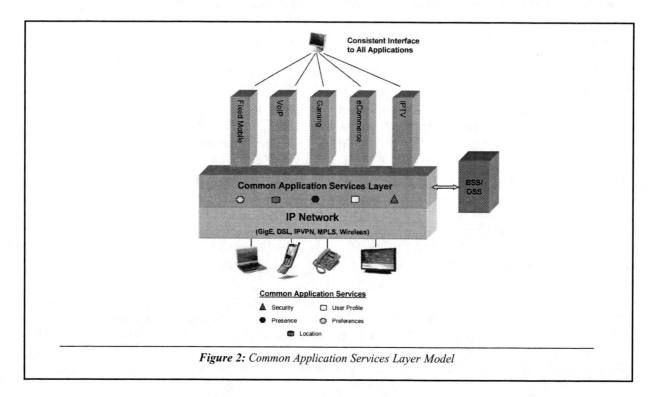

Figure 2: Common Application Services Layer Model

In addition to overcoming the deficiencies of the silo application model, a common services layer model has the following advantages:

- It allows service providers to offer a consistent set of services to end users. Users can access common application services such as presence and availability, location, directory services, subscriber profiles and preferences, and charging functions such as an e-wallet. Customers could enhance their experience by invoking different compatible application services. In addition, users could share common data such as authentication, preferences, and subscription/billing information across multiple applications of customer-specific data across services.
- It offers providers life-cycle cost savings and a faster time to market for application services. Providers can minimize their investment by implementing new services on a trial basis to determine their market acceptance and potential return on investment. In addition, a common services layer allows companies to differentiate application services from their competition. This differentiation could include the user interface, feature interaction across services, or even simplification of complex services.
- It fits in with the next-generation architectures being designed today to leverage service-oriented architecture (SOA) concepts, where each common service module provides a set of services to be invoked by other modules that, in turn, offer their own services.
- It allows third-party service providers to focus on developing specific application services using common services as needed without the cost of their development and support. Third-party applications can appear integrated into the company's own service model as just another one of its service offerings.

So far, we have discussed discrete application services and the underlying application architecture model that can better deliver those services, specifically a common services layer between the applications themselves and the network over which those services are delivered. Another consideration to address is real-time and near-real-time services and their historically differentiated services, architectures, and protocols.

Two Classes of Service: SIP and Web Services

Services delivered over today's networks fall into two distinct classes of services—real time (minimal delay commonly referred to as latency) and near real time (not sensitive to some latency). Real-time interactive services by their very nature require very low latency and very high reliability, with voice telephony being the most common example. For instance, VoIP service cannot typically maintain speech clarity if round-trip latency exceeds 200 ms. On the flip side, near-real-time application services are much more tolerant of latency because packet delivery and packet loss are not time-critical. Data applications delivered over networks and downloading music or video content to a storage device are examples of near-real-time services.

In an IP–based network (wireline or wireless), real-time application services primarily use SIP, whereas near-real-time applications are supported by Web services technologies such as extensible markup language (XML), simple object access protocol (SOAP), and Web services description language (WSDL). Until recently, SIP–based services were offered over IP networks that service providers designed and deployed with no unifying core infrastructure common to all. Today, a specification exists for such a core infrastructure for SIP–based services. It is called IMS, as specified by the third generation partnership project (3GPP). Unfortunately, no such unified core infrastructure has been specified for Web services, although several vendors offer proprietary solutions and promote them as a standard.

What Is IMS?

A lot of hype currently surrounds IMS. The specifications are still evolving and, in some areas, are minimal, as in the case of the service capability interaction manager (SCIM).

IMS deals primarily with the connectivity network infrastructure and uses SIP at its core. Although IMS was originally conceived as a service infrastructure to deliver multiple SIP–based services over third-generation cellular networks, it is now being considered as a service platform for both next-generation wireline and wireless networks.

Although the IMS core architecture is complex by definition, *Figure 3* illustrates a simplified version of the architecture. Over the past few years, industry momentum toward IMS has increased and service providers are starting to view it as a unifying service infrastructure capable of supporting IP applications in general—both SIP–based and Web services–based. Several major equipment vendors, including Alcatel, Ericsson, Lucent, Nortel, and Siemens, have embraced this vision and are developing IMS–compliant products, with general availability expected in 2007. According to the ABI Research study "IMS Core Networks for Fixed and Mobile Operators," network operators implementing IMS architectures will generate $49.6 billion in service revenue from IMS–based applications by 2011.[1] For wireline operators, most of this new revenue will originate from rich voice services, whereas mobile operators will benefit from new and enhanced services such as push-to-talk, interactive gaming, Web browsing, streaming content, and instant multimedia messaging. The report also predicts that fixed and mobile network operators will invest $10.1 billion in IMS infrastructure over the next five years.

The IMS specification talks about a session as a generalization of the familiar call session of the public switched telephone network (PSTN). The way the IMS specification uses the term, a session denotes a time interval during which one or more application services and users can be invoked and managed at the same time. The function in IMS that performs this common session management is called the serving call session control function (S–CSCF).

One key component of IMS is the home subscriber service (HSS) master database. The HSS, which contains customer subscription information and subscribed services, forms a logically centralized repository accessible to other IMS network components. This logical centralization of key data can support other capabilities such as authentication, subscription management, service policies, presence, and location, which can be reused across multiple applications in the IMS environment, similar to the common services layer previously described.

From the IMS standpoint, converged services means that both real-time and near-real-time services can exist in the same session. But how does IMS address that question? Unfortunately, the specification has only minimally addressed this capability by adopting the OSA Parlay and Parlay X gateway at

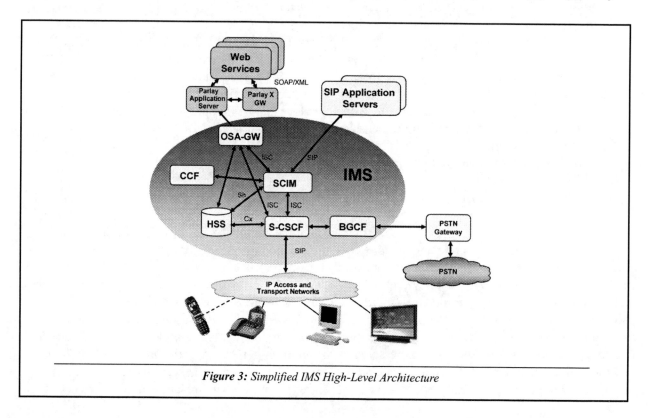

Figure 3: Simplified IMS High-Level Architecture

the application level. These were added to the core IMS architecture primarily to allow non-trusted third-party Web service providers to offer application services that leverage IMS components for their delivery. This level of interfacing between SIP–based services and Web services is insufficient for a tightly integrated converged SDP. To truly provide robust support for both service classes, a solution beyond the current IMS specification is required.

IMS also defines a SCIM component that is intended to provide a brokering function across applications and manage feature interactions as these different applications are invoked during a session. The specification language regarding the SCIM is minimal at best, so how a SCIM function actually operates and what capabilities it contains are really left for individual IMS implementations. As major vendors announce their IMS products, each addresses this service-brokering component in its own way.

The IMS specification today is extensive, and much has been written about its potential use and deployment. However, the scope of this paper is not to explain the IMS architectural framework in detail, but rather to discuss how its features can contribute to a converged services delivery platform solution. The IMS specification directly addresses the delivery of SIP–based services; how to merge that with support for Web services is the underlying issue.

Is There a Service Delivery Platform for Web Services?

Just as IMS is primarily specified for supporting SIP–based services, a complementary set of capabilities needs to be developed to support Web services using the same core system concept used for IMS. Moreover, the architectural framework for this core system should incorporate new information technology (IT) concepts such as SOA and Web services protocols. Today, the term SDP is often used to refer to this core system for Web services delivery. The concept of an SDP is widespread among service providers. Enterprise IT departments are also considering SDPs as a way to deliver both generally available and proprietary application services to employees and clients. However, despite the fact that the term SDP is widely used, no indus-

try standard exists for what an SDP comprises. Many vendors tout their specific solutions as an SDP, but most of these tend to center on a particular vendor's core products and competencies. To date, there does not seem to be a single solution in the marketplace that addresses everyone's concept of what an SDP should be.

However, it is still important to define what capabilities a basic SDP for Web services should incorporate. As discussed, Web services tend to be near-real-time applications, so the architectural framework is not as demanding as an IMS core infrastructure. But a Web service platform should include most of the same basic service delivery capabilities and common services available within a SIP–based services platform. These include service orchestration, Web session management, security, directory services, user profile management, presence, location, and notification.

An open interface to an IMS environment to invoke IMS capabilities such as third-party call control is also needed. This interface could be reached either indirectly through an OSA Parlay X gateway or directly through a SCIM interface. The drawback to using a Parlay X gateway from a Web services application is that two additional layers of interface are required. Parlay X currently has only 14 defined parts (functional areas) and is therefore restricted in its functional capabilities. Moreover, Parlay X is a gateway layer that depends on an underlying Parlay server to provide the actual functions in the IMS domain. Thus, any required IMS capabilities not addressed by the Parlay X gateway would have to be created using Parlay directly, a much more esoteric development environment.

The following is an example of a Web service that invokes a supporting SIP–based service. A user invokes a click-to-dial function on a name in his or her address book. Opening and managing an address book is a Web service that uses the Web services platform for execution. The click-to-dial action triggers the Web service platform to invoke a call control application programming interface (API) on the IMS gateway, resulting in a call being set up between the caller's phone and the party's phone whose number is extracted from the address book. As with all of the user's service sessions, this calling session is logged, in this case by the IMS platform, and a subsequent calling event is reported to an external OSS.

This example assumes that the user is initially consuming a Web services application (the address book) when the SIP/IMS application service is invoked. This fairly straightforward example is not too difficult to implement as we have described it. However, the inverse scenario is not possible, since the Parlay X gateway is one way and does not support the IMS core invoking a Web service during a session. Consider a user in the middle of a multimedia conference session on the SIP/IMS side pulling content from a Web service during the conference call. Unless the IMS conferencing application server has a direct Web services interface to the Web services platform, the content invocation during the session may not be handled by the architecture. In addition, if a conference participant wanted to invoke another Web services application, (e.g., open an IM session, share photos or video with another conference participant), that person cannot easily invoke that application directly from the IMS session. This unidirectional flow is one of the deficiencies in the IMS specification that needs to be resolved.

Just like the call session control function (CSCF) and SCIM together provide SIP--based session management in an IMS core platform, the session management function provided by the Web service delivery platform manages only Web sessions. It does not support management of sessions that have both IMS and Web services components. Going forward, session management in general needs to be enhanced substantially to support feature interaction for converged services. Although the IMS specification does not address the issue directly, the IMS SCIM component could provide such inter-domain session management. *Figure 4* illustrates a Web services delivery platform high-level architecture showing Web services interfaces (e.g., XML, SOAP, WSDL) to the IMS domain through Parlay X or directly to a Web-enabled SCIM. It also depicts Web service interfaces to the external OSS domains.

A Converged Services Delivery Platform

In the preceding sections, we discussed service delivery platforms for SIP-based (IMS) and Web services (SDP), respectively. We briefly talked about the IMS architecture specification and its capabilities to support SIP–based application services. We also suggested that a similar common platform and associated services would be required to deliver Web services, although no industry specification for one exists today. But how do we create a single, unified architecture model capable of sup-

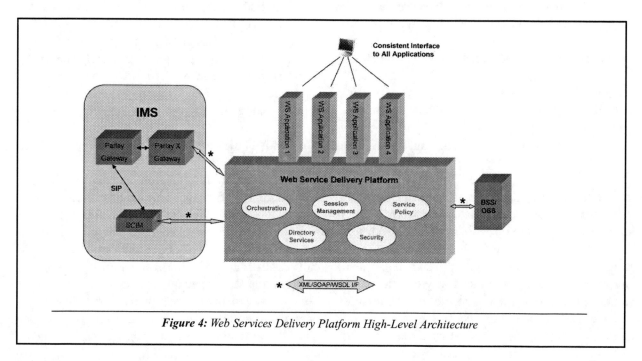

Figure 4: Web Services Delivery Platform High-Level Architecture

porting both types of quad-play services in a converged services delivery platform? *Figure 5* provides some insight into what such a model might look like. It encompasses support for real-time application services provided by SIP/IMS as well as near-real-time Web services provided by a Web SDP. Our earlier discussion of architecture models advocated a common services layer model that can be extended and incorporated into a converged SDP and made available to all supported applications. This common service components concept is also carried forward into our model.

Going forward, application services within both classes need to integrate and share their features during common service sessions. For example, SIP–based voice services should be able to invoke Web services such as presence, location, contact lists, and other information-sharing services. Even SIP–based call routing will need to invoke Web services to reflect end-user preferences and configuration settings. Conversely, Web services such as interactive single or multi-party gaming or video-conferencing need to support SIP–based services such as interactive multi-party communication. With this converged services concept in mind, our model needs to support both real-time and near-real-time services and provide integration and convergence capabilities within and across these mixed

service classes. This means that the platform could support both SIP and Web service applications. The convergence of these two technologies contained and integrated in a common converged services delivery platform would enable service feature interaction and standardized application behaviors across both service classes.

Figure 6 depicts the various functional components of a converged SDP and how they might interact. Note that all of the applications that support customer services, whether provided by the service provider or a third party, are in the upper layer and are capable of accessing all of the common components provided in addition to their specific core platform. The common components depicted here are only representative of a fuller set of services that could be available to all applications. What is important is that each of these common services is autonomous and not dependent on other services for its functionality. Both the IMS and SDP domains would access and leverage these core services to deliver their services.

But how can services be brokered and orchestrated in this converged services environment? As mentioned in our discussion of IMS, the SCIM could serve as a brokering function across multiple SIP applications. However, since the specification for

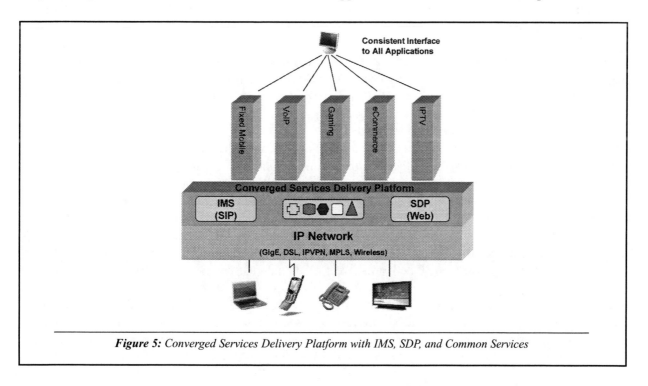

Figure 5: *Converged Services Delivery Platform with IMS, SDP, and Common Services*

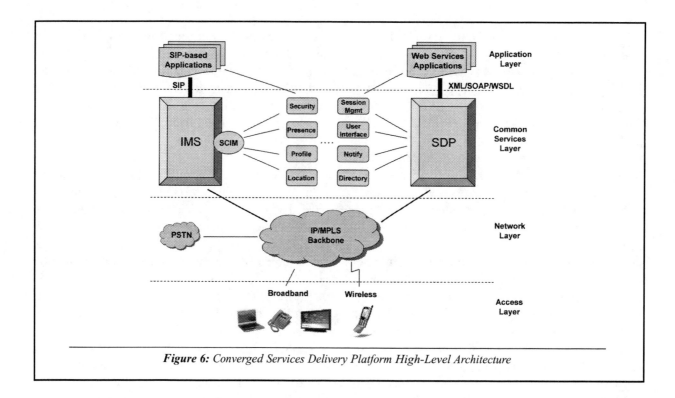

Figure 6: Converged Services Delivery Platform High-Level Architecture

the SCIM is minimal and does not define this, a specific IMS implementation would need to address this functionality. For Web services, modules in the SDP could perform the service brokering and orchestration across services. The orchestration of services leveraging these common components or invoking services in another domain could be accomplished using a session management component that coordinates and integrates both. This entire converged services environment would sit above the service provider's transport and access networks and be agnostic to both in delivering its services. This distancing, if you will, of the underlying network is crucial to be able to add new services in the converged services environment without regard to how or where they are delivered. This also permits end-user devices to evolve in their abilities and leverage multiple network topologies.

Summary

So far, we have addressed the new quad-play services enabled over next-generation IP networks. We have stressed that these new converged services need to be delivered to consumers seamlessly, regardless of how or what device is used. We discussed the two major functional architecture models that can deliver these application services—a silo model or a common services layer model—and pointed out the advantages to using the latter. We then described the IMS specification for supporting SIP–based services and discussed the concept of an SDP for Web services, pointing out the gap that exists in supporting user sessions that invoke both SIP-based and Web services (i.e., converged services). We proposed the need for a converged SDP that incorporates the capabilities of the two distinct platforms and provides a common layer of services that both IMS and Web services can use to deliver their respective services, provide common session management, and support seamless feature interactions in a session.

Such a converged SDP that can support the different elements of a quad-play bundle does not exist today. A service provider cannot buy one off the shelf from any single vendor. Certain pieces of the solution can be purchased, and several vendors currently sell an IMS core infrastructure product. A Web SDP can also be purchased from a vendor, albeit not to any common standard. The missing pieces are the converged services required by both platforms and the integration of those two solutions into a SDP to support combined services in a consistent and seamless fashion, regardless of what user device is employed to invoke the services.

Service providers will need to make sizable investments in development and integration to provide these services to their users.

It is unrealistic to think that a converged SDP can be built all at once. This is not feasible either financially or technically. Instead, we suggest such a platform needs to be developed over time with identified common components prioritized based on what services are required when. This means that service providers need to build a product road map that targets services they believe end users will want and buy. Such a product road map needs to be followed closely with architectural and functional system road maps that can deliver services in a cost-effective way that meet time-to-market goals.

The most productive approach would be to define a subset of the common services needed to support near-term next-generation services as part of a quad-play bundle and then progressively build on that subset as more services are identified. This requires service providers to identify what services they want to offer, their time frame for bringing them to market, and what underlying capabilities are required to support the services in their product road map. This would allow providers to build successful architectural and functional system road maps to create a converged SDP and optimize their investment.

Creating road maps and implementing a converged delivery platform for quad-play services require broad industry knowledge and experience, a sense of where the market is headed, and the necessary technology to stay competitive. Although no one can predict what new services might be successful in the marketplace, a converged SDP with reusable components capable of supporting common features would allow telecom service providers to remain competitive in an ever-evolving, fast-paced market.

Reference

1. ABI Research, "IMS Core Networks for Fixed and Mobile Operators," 2006, www.abiresearch.com/products/market_research/Mobile_Operator_IMS_Core_Networks.

Seeking a Common Vision of the Converged Home

Heather Kirksey

Senior Manager, Standards and Emerging Technologies
Motive, Inc.

As triple play gives way to quad play in the business plans of many service providers, mobile services are joining the already complex mix of voice, video, and data.

For service providers, the prospect of multiple new, potentially transient devices suddenly becoming at least part-time members of the home network may be daunting, as both the devices themselves and their impact on the home network undoubtedly will cause problems and drive support calls.

The devices, for example, need to be appropriately configured so that they can securely join the home network (given the difficulty many users have getting wireless fidelity [Wi-Fi] settings configured correctly for other devices, it seems overly optimistic to hope that they will not have similar problems with mobile handsets). To avoid degrading call quality in a congested home network, quality of service (QoS) priorities or other settings may need to be adjusted on other devices (such as the residential gateway) to ensure the best service.

If TR-069–based remote management has proven to be one of the key strategies in deploying services and reducing support costs for triple-play services in the home network, it makes sense that service providers would want to apply similar remote management capabilities to their mobile and converged offerings. Fortunately, the mobile industry has already begun work in the area of mobile device management through the Open Mobile Alliance (OMA). OMA was founded in 2002 to address the standardization needs of the growing mobility market, and this group has produced key specifications in the management of mobile or converged devices, including the OMA device management (DM) specifications, the OMA client provisioning (CP) specifications, OMA management objects, and download over-the-air (OTA). Viewed in aggregate, they represent the mobile industry's version of the TR-069 management specification family, intended to provide "setting initial configuration, subsequent updates of persistent information, retrieval of management information, and processing events and alarms" capabilities.

Much in these specifications is similar to the work that has gone into the DSL Forum's TR-069 standard. Many of the design goals are similar: agnosticism to the bearer network (Global System for Mobile Communications [GSM] versus code division multiple access [CDMA]), which is similar to TR-069's access medium agnosticism; a separation between the management protocol and the management objects, which is similar to the delineation between the CWMP protocol and the various device and service data models in TR-069; and the use of extensible markup language (XML) for the standardized message exchange. As in TR-069, management sessions do not depend on a persistent connection between the management server and the device, and management capabilities are bi-directional in that the management server can trigger a management session at any given time or the device can initiate a management session based on its own policies.

On the other hand, there are some key differences that generally reflect the evolution of the mobile industry in contrast to digital subscriber line's (DSL's) evolution. Compared to the DSL Forum's 12-year reign as the main standards body for that industry, OMA integrated seven wireless-oriented organizations into its charger in 2003. The long-term effect on standards consolidation will be positive, but the specifications currently seem to reflect this initial fractured nature, with multiple provisioning schemes described and interrelated pieces of work proceeding along somewhat separate paths (firmware download and device management are described in separate sets of documents, for example).

Many of the differences can be attributed to the difference in distribution models and problems to be solved. OMA–DM is based on a protocol called SynchML, which is used for both device management and for content synchronization. In the home network, data synchronization is hardly a pressing problem. But because the primary application of most smartphones is real-time access to e-mail, calendar, and address book, this capability is one of the most important in the mobile space. As a result, many capabilities and optimizations of the protocol are being designed to address this need rather than management.

The implications of these distribution models are apparent. Unlike the widespread self-installation model used in the DSL industry, most cell phones are procured by the consumer in person through authorized retail outlets, or by the IT departments of medium to large enterprises, which then distributes to employees. These distribution channels affect the provisioning assumptions: phones can often be partially or fully pre-provisioned at the point of sale before a consumer ever touches them. Enterprise IT organizations may want to control some of the device settings, leading to a shared management agreement between the mobile service provider and the company, which is atypical in the world of home networking devices. The mobile management infrastructure needs to be able to accommodate such needs, and service policies must account for enterprise personnel owning and managing devices that will function as part of the home network environment.

The demands of the mobile world also put different constraints on OMA work than those required for broadband devices. For example, the user interface (UI) of most residential gateways is an afterthought, most used by advanced technophiles to tune device configuration. A smartphone, however, is almost all UI, and users expect to be able to interact with it as well as control it. Mobile phone users might expect to be warned before configuration changes or expect to interact with the management application. Providing a means to prompt user input or display UI notifications is a real need in that market, but unimportant in the wireline industry. Similarly, although broadband device management should not be excessively chatty, the available pipe for management functions is much greater than for OTA functions. The demands of large bandwidth operations such as firmware or other image downloads on the cellular network, are much different from similar operations across copper or Ethernet and must be considered accordingly.

Probably the biggest difference between the TR-069 specifications and the OMA–DM documents is the set of available managed objects. In the gateway, voice over Internet protocol (VoIP) and IP set-top box (STB) data models, the DSL Forum has specified a robust set of parameters that include almost all imaginable pieces of readable or writeable device state. The standardized objects in OMA currently only include baseline information required to communicate with the DM server, and basic network configuration and device properties such as manufacturer, model, and firmware version. Their intention is for device manufacturers to "provide servers with the necessary information they must have in order to manage" their devices by "publish[ing] descriptions of their devices as they enter the market." Relying on margin-obsessed device manufacturers to spend the development effort to identify, make available to management, and publish the features has not led to a large number of capabilities being exposed.

Additionally, handset manufacturers are unlikely to expose additional functionality in the same objects or parameters, even if for similar functions or providing similar statistics. This requires server vendors to do far more work to understand the capabilities and semantics of the models provided by different vendors, a fact that may hinder operators looking to provide management with the same level granular control they are used to with customer-premises equipment (CPE).

Understanding the points of standards commonality and differences, as well as their drivers, is an important first step in moving mobile and home networking domains ever closer. Beyond that, however, the operators that find themselves increasingly responsible for both would be well served to ensure that management standards are advancing in common directions. Although some technical details may never converge, as the needs of managing an OTA device are different from those of managing broadband CPE. At the least it is necessary to ensure that the combination of available standards is adequate to serve the needs of quad-play deployment. A common vision of the converged home across the quad-play ecosystem will best serve the service providers, the vendor communities and, most important, the consumer.

Quad Play Requires a Universal Service Delivery Platform

A Universal Service Delivery Platform Enables Telco, Satellite, and Cable Operators to Deploy, Brand, and Bill a Rapidly Expanding Universe of Content and Services at Lowest Total Cost of Ownership

Venkat Krishnan

Director, IPTV Solutions
SeaChange International

In recent years a lot of attention has focused on communication protocols as the key to quadruple play—the ability to blend content and programming across a branded offering that combines TV, phone, Internet, and wireless services. The conventional wisdom states that the Internet protocol multimedia subsystem (IMS) must be fully deployed for consumers to enjoy the quad-play experience. Once diverse access types (e.g., IP television [IPTV], digital subscriber line [DSL], wide-area network [WAN], general packet radio service [GPRS]) can interoperate, then diverse services can converge.

But if the promise of any content, on any service, on any network, on any device is to happen, then the language of diverse services is not the only thing that must become universal—the service delivery platform (SDP) on which services run must as well.

In fact, the platform comes first—specifically the platform most in demand now: IPTV. IPTV is the next-generation service that is the most technology-intensive and offers the most eye appeal to consumers—and it is fully deployable today. If it places the right bet here, an operator will enjoy both a short- and a long-term advantage.

In the short term, operators can generate revenue from the on-demand options that consumers want now. In the long term, they can add even more quad-play options as IMS arrives without ripping out existing infrastructure, operating multiple technology stacks, or prematurely investing in technologies not yet ready for prime time.

The fact is, at some point every operator makes a platform decision about how to deploy services—quad-play or otherwise—even if by default. That decision either ignores the difference between a service and the delivery platform on which it runs or exploits the difference to increase scalability, flexibility, performance, and other benefits.

To ignore the difference means that every service brings its own client, its own navigation, its own way to capture user preferences, its own business rules engine, and so forth. To exploit the difference means that diverse services can share common functions in a middleware layer. Not only does this reduce duplication and operations costs, it also allows operators to plug and play best-of-breed functions such as digital rights management (DRM), set-top boxes (STBs), and video servers into services where it makes logical sense. Even where operators do not want to replace functions, they may still wish for functions to be decoupled— for example, user entitlement management from content asset management—creating new opportunities for market differentiation.

Nowhere are these opportunities more evident than in a multi-business operator model. In this model, larger operators sell services to smaller operators, which resell those services under their own brands to consumers. This model works only if the services are truly differentiated. Operators cannot simply resell the same programming packages, the same business rules, and the same look and feel—in other words, the same brand experience—as their competitors. But if functions are decoupled, they do not have to. How they tailor one function does not limit how they tailor others. That gives them and their customers what they demand most, both in IPTV and in quad play—ultimate choice.

What Consumers Demand – Ultimate Choice

Because choice is so important to the consumer, it sets the ground rules for the operator, and therefore the technology.

Consumers want choice of both content (e.g., movies, games, music) and context (e.g., content delivered when they want, where they want, in the form they want—personal video recorder [PVR], digital video recorder [DVR], video on demand [VoD], subscription VoD [SVoD]—live, prerecorded, on-demand, or interactive). They want their TV on the Web, the Web on their TV, videos on their wireless phones, and e-mail and text messaging on everything. They also want to set content and context preferences in whatever way works for them— for example, via an STB, a personal computer (PC), or a mobile phone. In addition, different members of the same household will express different preferences.

Helping consumers control context and content is a great example of where operators can make money from a platform-enabled quad-play lifestyle—provided the platform has features such as the following:

- Preference engine—For setting content and context
- Recommendation engine—For alerting consumers to content and context preferences they might enjoy based on previous choices
- Search engine—For finding content from "the long tail," i.e., all the movies, home videos, photos, network programs, etc., available regardless of location

Some consumers may in fact be satisfied to watch only scheduled network programs on an analog TV set and never wish to pause the action to check their e-mail. What counts is the freedom to indulge in as many preferences as possible in ways that create meaningful brand differences.

What Operators Need – A Difference

If choice defines the consumer, difference defines the operator. Difference is how you stand out, how you avoid competing on price, and how you brand. It is a reflection of the choices you offer your target consumer—for example, either by marketing a

Consumers demand:	Operators need:
ß Content whenever and wherever	ß Reliable on-demand system
ß Flexibility to tailor and manage services on their own	ß Rich portfolio of on-demand applications
ß More interactive experiences (PVR/DVR)	ß On-demand application deployment
ß More on-demand experiences (VoD/SVoD)	ß Low costs
ß Affordable prices	ß State-of-the-art functionality and performance
ß State-of-the-art function and performance (e.g., HDTV)	

Table 1: Consumer Demands Drive Operator Needs

greater abundance of choices in more accessible ways or by marketing a more targeted set of choices to a specific demographic.

Either strategy creates differentiation from core product attributes—those outlined on the right side of *Table 1*. But sustaining those attributes—keeping applications exciting at an affordable cost over time—is a challenge. What makes an offer different today will not make it different tomorrow. To sustain product differentiation over the long run operators need the following key success factors working in their favor:

- Time to market—Operators need to bring new service options to the consumer at least as fast as competitors do. The challenge: unlike in previous eras, when services were functional "stovepipes," services today cannot be considered deployed until they are blended with other services, which can take time. Furthermore this blending occurs at multiple points—for example, in the collection of billing data and the presentation of client navigation. Users will not tolerate the equivalent of yet another TV remote on the coffee table every time something new is added to the service portfolio. Marketers meanwhile will want the opportunity to invent new products that combine features of various quad-play services in interesting and creative ways.
- Revenue growth with reduced customer churn—Top-line growth requires keeping existing customers happy while attracting new ones. At times, these two objectives may seem in opposition. Existing customers expect continuity with what attracted them to the operator in the first place, while both existing and new customers are attracted by new offers that are better than what they have now. The optimum solution is to build on what works today—such as VoD and games on demand—in a way that is easily extensible tomorrow.
- Low deployment risk—Deployments grow and become more complex as new services, applications, product configurations, and customers are added. Growth and complexity increase risk. There are more things to go wrong and ways for things to go wrong. In addition, as services become less stovepiped and more interdependent, a problem in one place is more likely to cascade to other places. Scalability of the solution is key. Deployments that start small should be able to grow quickly yet remain stable as operators gain customers and experience. You increase scalability when you componentized functions as discrete modules with clean, well-planned interfaces that act like fire doors that close when bad data tries to pass through. Isolating discrete functions also makes it easier to pinpoint an issue or swap out a faulty component without impacting working parts of the system.
- Low total cost of ownership—"Our costs are less so we can sell for less"—a refrain often heard at auto dealerships—also applies to quad-play operators. The lower the costs, the more attractive the prices and the more money left over to deploy new services. Lowering deployment risk reduces cost of ownership—another reason to implement easily extensible modular systems. Modularity also allows operators to swap in best-of-breed components—components that may be less expensive to buy and more cost-efficient to use. The STB is a great example, accounting for 70 to 80 percent of operator deployment costs—with most of that coming from proprietary client functionality buried within silicon and C code. These costs can drop substantially to the extent that the following becomes true:
  ○ STBs are interchangeable commodities
  ○ Clients are written in a portable language such as Java, JavaScript, or hypertext transfer protocol (HTML) that can be easily modified
  ○ Performance-optimized functions are server-based

Another savings opportunity is the effort needed to integrate services with an operator's existing business support system (BSS)/operator support system (OSS). The more seamless the integration, the faster and less costly a service is to implement and maintain over time. And to the extent that any service is able to inherit these types of cost savings, the greater the opportunity to keep offering competitive services at competitive prices.

It Is All about the Architecture

Although these are four success factors—speed to market, top-line growth, low deployment risk, and low total cost of ownership—they have a lot in common, a fact that points to a single underlying strength. The following is what they have in common:

- Operator success factors are mutually reinforcing—It is no accident that an operator that knows how to speed new services to market will also likely experience healthy top-line growth, or that an operator with low deployment risk will enjoy low cost of ownership. Risk increases cost. And slow deployments will likely turn off both existing consumers and miss windows of opportunity for attracting new ones.
- Operator success factors exploit common technology features—Success factors are mutually reinforcing to the extent that common technical attributes (like the ability to plug and play functions easily) speed services to market, reduce risk, and lower cost of ownership, which again spurs top-line growth either by attracting new business or reducing customer churn.
- Operator success factors are architecture-dependent—What makes these similar features alike is architecture (i.e., a conscious strategy to map components into services based on cost, performance, and best-of-breed functionality). For example, to deploy the user interface of an electronic program guide as a discrete function allows users to swap out commodity STBs that cost less and perform better.
- Operator success factors are service-agnostic—Whether the service is IPTV, mobile telephony, Web browsing, or some combination, the distinct value of architectural advantages is that they tend to apply everywhere. They can bear fruit again and again over different services and over time. Take the example of a modular BSS/OSS adaptor. Once an adaptor module exists for easy integration with a specific BSS/OSS, it does not have to be reinvented for every service or application. That speeds deployment, reduces risk, and reduces total cost of ownership all at the same time—for any service or application that needs to talk to the BSS/OSS—all of which attracts new customers and helps satisfy existing ones with new offers.

Convergence on Trial

What this shows is that architecture, done right, provides a powerful single point of leverage to keep operator offerings fresh across services, applications, and content. Creating brand differences, however, takes more than just a fast, low-cost, low-risk, sustainable way to refresh services. You also have to tailor services as differentiating user experiences that stand out in the consumer's mind. That is why some device makers are doing limited rollouts of convergent services (e.g., programmable video recording via mobile). They are hyping these trials as harbingers of more ambitious hybrids to come. But operators are not where device makers want them to be. Rather than architect deployments around a device, operators would rather deploy devices around a deployment architecture so they can entertain all branding possibilities.

As *Table 2* shows, operators have an almost unlimited number of ways to tailor their offerings. Consider that there are four basic services that define quad play, each of which consists of a long list of elements. And each of those can be tailored differently depending on the service, user preferences, cost, and other factors. In some respects, navigating a cell phone menu will be different from (and the same as in others) navigating a TV's electronic program guide. Good branding would call for elements such as color, company logos, and page layout to be consistent. But content sources would probably be different, and presentation style could also change depending on whether the presented information were on-demand TV listings, unread e-mails, or something else.

Content branding calls for access to specialized services and tools, such as those that help the following:

- Design electronic program guides (EPGs)
- Add logos, colors, and backgrounds to set-top clients
- Design backgrounds, slideshows, wall of posters, and other user interfaces
- Acquire and aggregate content
- Compile and promote service bundles
- Create Moving Pictures Experts Group (MPEG) assets with multiple video back-

		Services			
		TV	**Phone**	**Web**	**Mobile**
Elements	**Content**	• On-demand • Interactive • Convergent	• Interactive • Convergent	• On-demand • Interactive • Convergent	• On-demand • Interactive • Convergent
	Applications	• On-demand • Interactive • Convergent	• Interactive • Convergent	• On-demand • Interactive • Convergent	• On-demand • Interactive • Convergent
	Hardware	• On-demand • Interactive • Convergent	• Interactive • Convergent	• On-demand • Interactive • Convergent	• On-demand • Interactive • Convergent
	Business rules	• On-demand • Interactive • Convergent	• Interactive • Convergent	• On-demand • Interactive • Convergent	• On-demand • Interactive • Convergent
	Navigation	• On-demand • Interactive • Convergent	• Interactive • Convergent	• On-demand • Interactive • Convergent	• On-demand • Interactive • Convergent
	Entitlement	• On-demand • Interactive • Convergent	• Interactive • Convergent	• On-demand • Interactive • Convergent	• On-demand • Interactive • Convergent
	Promotions	• On-demand • Interactive • Convergent	• Interactive • Convergent	• On-demand • Interactive • Convergent	• On-demand • Interactive • Convergent
	Other	• On-demand • Interactive • Convergent	• Interactive • Convergent	• On-demand • Interactive • Convergent	• On-demand • Interactive • Convergent

On-demand examples: VoD, games on demand, broadcast
Interactive examples: Red button apps, voting, e-commerce
Convergent examples: Caller ID, video blogging, home media programming, PVR via mobile

Table 2: When Tailoring an Offer, an Operator Should Entertain All Combinations of Services, Service Elements, and Service Types (on-Demand, Interactive, or Convergent)

grounds and still frames, as well as video and audio elements

Non-content branding elements would include entitlements and business rules, functions that determine, respectively, which content or applications a user is entitled to consume and under what conditions. Entitlements may apply one way when the content is a premium on-demand movie the user has rented, which the business rules say can be viewed up to a certain number of hours after payment. But entitlements and business rules may apply a different way when the content is a music video the user has purchased for download off a music service.

Added to this mix of service-versus-element is another layer of complexity, that is, service-versus-element-versus type of service (i.e., on-demand, interactive, convergent). In a convergent system, for example, on-demand movies might be played on either a high-definition TV (HDTV) home theater or on a wireless phone, among other possibilities. Again, various entitlements, business rules, and other elements will apply differently depending on the case. Furthermore, if the movie has interactive content (e.g., alternate endings, premium tie-ins to a Web site), additional entitlements and business rules may also apply. For example, you may be able to chat with a star in the movie or insert your own image in a scene.

What is clear from this picture is that there is far more to keeping your offering fresh than just updating the various elements of content, applications, entitlements, and so forth. You also have to tailor these elements in sync with audience demands while reinforcing a clear brand distinction. As *Table 2* shows, there is an almost unlimited number of ways to do that—especially when convergent (IMS–enabled) applications become available.

Get Ready for IMS Now

In fact, when IMS–enabled applications do become more available, operators will jump into the tailoring game or quickly become commoditized. That is because IMS only has value in a blended environment—and to blend applications, you have to tailor them. After all, what is the point of allowing applications that are running on different services to blend if they do not in fact blend? And if they do blend, then there must be value propositions to support that effort. As these value propositions emerge, they will drive further blending, further tailoring, and even more value propositions, which will drive further tailoring and blending—and so on and so on.

This begs the question: Why wait for an IMS architecture to emerge? As *Table 2* shows, there are a lot of opportunities for mixing and matching that can be exploited today to differentiate service offerings. Why not jump in now and get a head start at building the expertise and infrastructure required to do that? Operators cannot make money selling convergent applications that do not yet exist. But they can make money selling a convergent experience. That is an experience where applications are as follows:

- Consistently branded across services
- Capable of invoking elements tailored to satisfy consumer choice and create brand differences
- Highly interactive
- Capable of satisfying on-demand requirements

Selling a convergent experience before IMS actually takes hold offers three big advantages: it gives the operator credibility in the marketplace as a forward-thinking provider; it gives the operator experience at running convergent applications so the transition to IMS will be smoother; and it implies you already have in place a framework for efficient quad-play deployment—where discrete service modules share a componentized backend infrastructure you don't have re-engineer every time consumers want something new.

IPTV: Giving Consumers What They Want

In fact, the only reason not to jump into the tailoring game now is if you did have to re-engineer your infrastructure every time consumers want something new. If your billing function only works with IPTV and not with mobile, it will be hard to add mobility to your IPTV offering. If you cannot stream TV to both an STB and a wireless personal digital assistant (PDA), that is another blended offering you cannot provide. When what you own is a collection of stovepipe applications, then the only way keep ownership costs down is to not re-engineer them. The problem is, what happens when a competitor in your market area can mix and match service elements as well as keep total cost of ownership down?

But, an operator might argue, what happens when IMS technology does arrive and convergent deployments do become widespread? Will we not have to upgrade then anyway? Will that not level the playing field? In other words: Does any infrastructure investment make sense now, given the risks of how future technologies might evolve?

But here is another question: What about IPTV? IPTV is a technology that is fully deployable now and is certain to be at the center of any future IMS–based offering. It is also the technology that most engages consumers today. IPTV already offers the following:

- High-definition digital-quality video
- Surround-sound stereo
- Mass-audience entertainment content
- On-demand delivery
- High interactivity
- Digital lifestyle features such as the following:
  - Stop live action

○ Time shifting
○ Electronic program guides
○ Program-in-program viewing
○ Camera selection

These features are likely to form the center of gravity for any new quad-play investment—pre– and post–IMS—precisely because they are the most sensory-engaging and therefore also the most technology-intensive. Also, IPTV demonstrates an important lesson about quad play: users, and therefore the services they want, often drive the timing of infrastructure investments. Witness the demand for expensive plasma screens, HDTV receivers, and most recently Blu-ray players. If operators are willing to wait until IMS standards completely sort out before they experience the IPTV difference, users are not.

That means to get quad play right, you had better get IPTV right first. If you are a cable operator, then IP video services are the base from which you will move into other services. As you go, you want to avoid ripping out existing infrastructure or operating multiple technology stacks. If you are a telco, you want to avoid that fate too, but in your case, rather than build off an existing IP video base, your task is to add IPTV to existing telephony and wireless. So in either scenario—building out or adding on—it is your choice of IPTV solution that will largely decide how "IMS–ready" your offerings are "out of the box." In other words, can you easily plug new services into existing ones and easily tailor and blend the services you already have?

But telcos have a greater stake in their IPTV decision than cable operators. That is because consumers already think of cable companies for TV—and because telcos with no TV customers have no TV consumers. Every telco IPTV customer must at some point be taken away from an entrenched TV competitor. Telcos must also ask consumers to change how they think about where to get TV—a much harder task than just winning consumers from other telcos.

That means the first time telcos offer their customers a convergent experience, it had better be great. And they have to do it without making mistakes, especially when it is still early—before they have had a chance to earn a good reputation. As the cable industry has discovered, consumers can be very unforgiving when it comes to their TV service.

IPTV as Universal Service Delivery Platform

So whether you are a cable provider or a telco, IPTV plays two roles—it delivers a branded TV offering and anticipates the quad-play experience of IMS.

This describes a service-oriented architecture (SOA) in that services are decoupled from each other as discrete modular components. In particular, back-office functions (such as those in *Table 3*) are decoupled from the end-user applications (e.g., EPG) they support. That makes all services easier to isolate, and therefore tailor, than when functions are hard-wired into an entire service.

Elements such as an STB or a DRM module are also easier to swap out. And any cost or performance benefits—derived either by tweaking a particular module or replacing it entirely—are inherited by all services that use that module. So rather than increase total cost of ownership or deployment risk, tailoring an element such as business rules—say, to work one way for movies and another way for games—can actually *decrease* total cost of ownership and deployment risk. Then when you go beyond IPTV—say, from movies and games to IM—you still only have one business rules engine to deal with. And it is the same rules engine with which your technicians are already familiar, which again drives down total cost of ownership, deployment risk, and time to market.

Five Common IPTV Architecture Mistakes

These examples show how IPTV, properly architected, facilitates quad-play deployment—exactly the type of deployment IMS anticipates. But the reverse is also true. Architect IPTV the wrong way and it becomes a huge quad-play (and therefore IMS) roadblock. Consider, for example, five common IPTV architecture mistakes: end-to-end, application-centric, organic, device-centric, and BSS/OSS–centric.

• End-to-End—End-to-end deployments view IPTV as essentially one closed monolithic sys-

IPTV factors for quad-play success	Will tend to be	And architecture-enabled by
• Speed to market • Revenue growth with reduced customer churn • Low deployment risk • Low total cost of ownership	• Mutually reinforcing • Based on common technology features • Architecture-dependent • Service-agnostic	• Functions componentized as discrete modules • Cooperation between modules via clean, well-planned interfaces with built-in safety features • Reusable adaptors for easy integration with external services such as BSS/OSS • Components deployed based on cost or performance

Major function	Functionality
Subscriber/account management	Device management, service allocation, provisioning, billing, preference and profiles, transaction history
Product offering/campaign management	Packaging, pricing, rating, promotions, and discounts
Digital rights management	Content, operator, and subscriber protection
Service and asset management	Channel lineup, preferences, pay-per-view management, service configuration, barker channel, application service management
Network and systems management	Video service assurance, alarms and monitoring, performance/video monitoring, administrative dashboards, ticketing and troubleshooting, content life-cycle management

Service-Oriented Architecture: Key Features

N-tier
Set-top box/service middleware/operator functions

Functional partitioning
Navigation/entitlements/asset management, etc.

Performance optimization
Performance-optimized streaming hardware/network-based QoS, etc.

Application decoupling
Open interfaces/standard protocols, etc.

Ready-to-market applications
Video on demand, TV mail, etc.

***Table 3:** Back-Office Functions Are Decoupled but Shared in Common by Quad Play Applications in a Services Oriented Architecture*

tem. Everything is hardwired to everything else, so changing a function here has the potential of breaking a function there. Also, functions are designed to work specifically with each other, i.e., they are tightly coupled. That means that all functions are only designed to work in a single context, i.e., on-demand video. Making the same function (e.g., navigation) support a different context (e.g., video downloads to wireless phones) might not be possible. Finally, both operators and users are locked into using vendor-supplied functions. All of the above make solutions hard to tailor or scale without compromising total cost of ownership and other quad-play success factors.

- Application-centric—This is when the entire IPTV offering is deployed as a software layer that rides on top of operating systems, middleware network services, and hardware. Application-centric architectures also tend to be end-to-end, with the single application trying to do everything (as opposed to an SOA, where application functions are deployed as discrete best-of-breed components). Typical application-centric examples come from PC software companies looking to move into the TV space. As such, they often lack the "industrial strength" expected of a true IPTV company with deep expertise and experience across the entire IPTV technology stack.

- Organic—Also called homegrown, this is where an operator builds out its own service architecture by stitching together services and functions as they are added to the deployment. The problem with organic solutions is that they are designed around the needs of one particular operator. They are not designed to transfer well to other operators. As a result, the alignment of functions versus modules may be inconsistent and unclear—interfaces may be ad hoc—and a lot of overlap may exist between modules. As with end-to-end, organic solutions do not scale well and tailoring is impossible without undermining quad-play success factors.

- Device-centric—This architecture is also very organic, except that now it is the device vendor (say, an STB maker) that stitches the modules together, rather than the operator. The architecture suffers from the same weaknesses as organic. It also suffers from the additional

weakness of being oriented around a particular component of the solution—i.e., the vendor's device—rather than the solution as a whole.

- BSS/OSS centric—A BSS/OSS is the infrastructure that enables telcos to create, deploy, manage, and maintain network-based services. A BSS/OSS–centric services architecture, however, is something of an oxymoron since the BSS/OSS itself is not fundamentally concerned with services. It is not designed to know about entitlements, business rules, client navigation, and the other shared service resources. It is designed to know about infrastructure resources such as QoS, switch configurations, firewalls, and dial tone. Organizing shared service resources within an SOA—on top of the BSS/OSS—avoids trading network priorities off against service priorities and vice versa. It also avoids the painful, time-consuming, and potentially destabilizing effects of re-engineering the infrastructure every time you want to add or tailor a service.

Deploying a Service-Oriented Architecture

Organizing around services—IPTV first, and then convergent quad play—is a layered process. The strategy is to pick the low-hanging IPTV fruit first. Give consumers a convergent experience they will enjoy now, but also nurture quad-play success factors as IMS builds out. That strategy relies on an SOA that decouples functions (that are simultaneously service-enabling and network-agnostic) as discrete functions decoupled from each other and also from the BSS/OSS that controls the larger infrastructure on top of which they run.

Those functions include business rules, STBs, clients, navigation, EPGs, entitlement, content asset management, ordering, and billing. Actually, the specific functions are not what are important since the whole point is to be able to plug in new functions (e.g., those supporting new quad-play applications) that might not have been anticipated when IPTV was installed.

Figure 1 shows this architecture in action as a universal SDP. The three layers are as follows:

- Client—This environment is where the user interacts with the world of on-demand, interac-

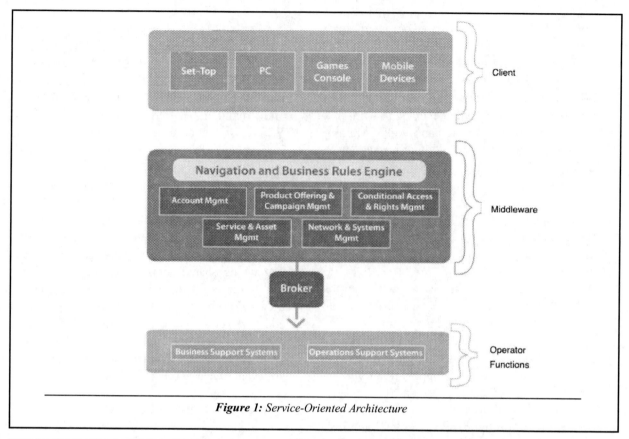

Figure 1: *Service-Oriented Architecture*

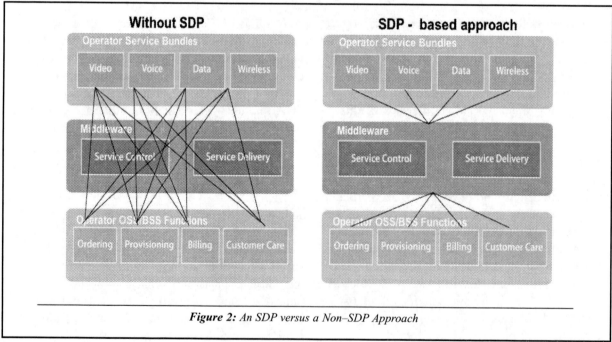

Figure 2: *An SDP versus a Non–SDP Approach*

tive, and convergent applications—everything from movies on demand to games to TV mail. As such, it is operating system-, STB–, and network-agnostic. Core technologies include third-party application launch and connectivity with back-office functions. To shed costs, clients may be deployed on devices that are both resource-constrained and sourced from multiple vendors providing diverse services supporting open standards such as Java, JavaScript, and HTML, and emerging standards such as open cable application platform (OCAP) and multimedia home platform (MHP).

- Back office—This is where application logic resides—both the logic shared by multiple services (e.g., billing or business rules) and application-specific logic (e.g., TV mail). Some functions (e.g., navigation) talk to client code. Some (e.g., business rules) do not talk to a client but do talk to other functions (e.g., conditional access, rights management). Some back-office functions also talk to BSS/OSS integration functions in Layer 3. For example, billing middleware that tracks movie rentals will talk to the billing system that actually invoices the customer.
- BSS/OSS integration layer—This layer provides invoicing, customer care, legacy data access, and other services external to the quad-play technology stack that support administration of the business rather than actual service delivery.

To get an idea of how this approach streamlines operations and deployment, look at *Figure 2*. In conventional deployments (left side), functions are not componentized, and components are not decoupled. Every piece of logic is hardwired to every other piece of logic that it needs to create an appli-

cation. This creates a "rat's nest" of one-to-one connections that make tailoring and updating virtually impossible while still providing low total cost of ownership and other success factors.

In a service-oriented deployment (right side) functions are componentized, and components are decoupled. This allows services to easily exploit common facilities without duplication or a lot of overhead (human or technology). On the client side, this allows fast, incremental deployment of best-of-breed applications into commodity hardware (see *Figure 3*). In the middleware layer, this allows creation and delivery of revenue-generating video and other quad-play bundles. And on the operational layer, this allows full integration with BSS/OSS, customer support, accounting, and other back-end systems through an open, flexible, scalable architecture without duplicating or replacing the functionality of those systems.

An SOA also enables easy transition to an IMS world. It allows service providers to gain experience with the type of blended distributed environment IMS envisions. Operators reuse functions such as device management and content delivery across applications now just as they will reuse them later with IMS, providing a high level of service continuity in the transition process. This is the

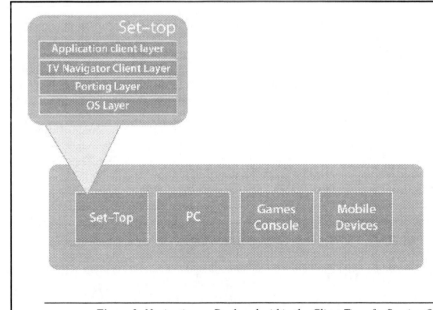

In an SOA, a single client navigation component can be reused (and easily tailored) for fast cost-effective deployment of multiple applications for TV and other video-enabled devices, including the following:

- Broadcast, on demand (VoD), and time-shifted TV
- EPG
- PVR and network PVR
- HDTV
- Games
- DVD on demand
- ETV
- Communication applications (e.g., mail, chat)

Figure 3: Navigation as Deployed within the Client Tier of a Service-Oriented Architecture

sweet spot of IPTV deployment: a great suite of IPTV applications and content in a plug-and-play architecture that accommodates choice, tailoring, and performance; keeps cost of ownership low; and protects your investment heading into the IMS future.

The Ultimate Test – Multi-Business Quad Play

Any operator can exploit the benefits of an SOA on which to deploy a universal SDP. That is because services are no longer defined by particular end-to-end collections of logic and rule sets. However you wish to define your business—whether by introducing new services and applications or by creative tailoring of the services and applications you already have—is up to you. Navigation, entitlements, business rules, content asset management, access control, and other core middleware components are now decoupled, so you are now free to use and reuse them differently for different applications and different kinds of content.

These advantages benefit one operator business model in particular, however. That is the multi-business operator. In this model, a large operator sells services (e.g., IPTV, Internet, IP phone, and wireless) to smaller operators. These other operators might be broadcasters, ISPs, content producers, telcos, cable companies, or whatever. They in turn resell these services under their own distinct brands to end users.

One branding element might be a different "skin" for STB navigation (i.e., different colors, graphics, information layout). Another might be more flexible business rules (e.g., longer "happy hours" or more liberal 2-for-1 movie rental promotions for just-released titles).

The key to making this model work is the ability to easily, rapidly, and cost-effectively tailor and blend services differently for different customer sets on a common platform to achieve economies of scale. Those economies disappear if operators have to implement redundant platforms on which to base services that look or behave differently. For exam-

ple, they would not want to implement redundant BSS/OSS connections (or possibly implement redundant BSS/OSSs). Nor would they want to install multiple IPTV systems just because its rules engine only works if you manage content assets a certain way (because rules and assets are tightly coupled). Nor would they want to install redundant IPTV back offices and gaming back offices (or a fill-in-the-blank back office) for the same reason.

The multi-business model is the ultimate test of whether your SDP is truly universal. If your platform can support this model (even if you do not happen to deploy that model yourself), then you know it can very likely support the business model you have now or are likely to have in the future. Whatever your business model, you have the power to differentiate multiple operations throughout the value chain all the way from content to final application deployment.

Choice Rules

As the multi-business model illustrates, the power of your technology base ultimately rests with the ability to satisfy consumer choices at will. Just as consumers now want video content on demand, in an IMS world they will want the next new hot application (whatever that may be) on demand as well. Not having a universal SDP will be like having your feet stuck in concrete.

A logical place to start is IPTV. Certain to be the centerpiece of any quad-play offering, the right IPTV architecture gives consumers the rich multimedia experience they want now and gives you both capability and the experience at offering a very convergent product well before IMS actually arrives.

Knowing what to do next is always difficult, especially when technology moves so quickly and so many technology investments can be make-or-break. But with so many great IPTV applications here today and so much hype about what IMS will mean tomorrow, standing still is not a choice.

Key Success Factors for Quadruple-Play Services

Erik Larsson

Vice President of Marketing, Netcentrex Converged IP Communications
Comverse

Introduction

Quadruple play is used in the telecommunications industry to describe a blend of voice, video, and data with mobile services. It offers a range of services in a seamless environment, which encourages subscribers to stay with a single service provider, thereby reducing churn and opening up new revenue-generation potential.

Analysts predict a strong market uptake for fixed-mobile convergence (FMC) and Internet protocol television (IPTV)—the key components of quad-play services. The Multimedia Research Group (MRG) forecasts that the number of IPTV subscribers worldwide will grow from 4.3 million in 2005 to 36.8 million in 2009. According to industry analyst Visiongain, FMC will drive fundamental change in both fixed and mobile industries and the market is set to grow to a value of $74 billion by 2009.

Quad play is quickly becoming established in the United States and Europe, as witnessed by a range of recent announcements (e.g., Sprint Nextel, Virgin Mobile, Neuf, Free, Orange, Telenet). Initial packages are offering cut-price deals to encourage customer uptake. The much bigger (and more interesting) challenge is to generate value from actual convergence between services.

This article describes the stages in which quad play services can be implemented, provides service examples, and gives a range of key success factors for each.

Quad play implementation stages are as follows:

- Service deployment—The initial service deployment, with a strong emphasis on attractive pricing of combined fixed-mobile service bundles and providing a single point of contact for sales, billing, and customer service
- Value-added and interactive services—More advanced applications and interactive services, including dual-mode handsets, FMC voice and video communication, and interactivity between IPTV and other devices
- Lifestyle services—Service bundles for communication, information, and entertainment, aimed at groups of users with similar interests and usage patterns

These stages do not necessarily represent phases in time, but rather a measure of the depth of quad-play services: a given service provider may choose to implement a mix of several stages at the same time.

Service Deployment

Overview
For consumers, the main advantage of the first stage is the convenience of receiving a single bill and a discount by getting all fixed and mobile communications services from one source. Service providers benefit from reduced customer churn and the potential to increase revenues and market share.

Examples

There has been a recent flurry of quad-play announcements in Europe and the United States. In France, the market has been particularly active since the second quarter of 2006 with the announcements of FMC services by all major operators: Orange, Free, and Neuf. France Telecom has combined its broadband, Internet services, IPTV, voice over IP (VoIP), and mobile services under the Orange brand, confirming the mega-trend toward integrated operators offering quad-play services.

In the United Kingdom, NTL Telewest has integrated Virgin Mobile to offer a quad-play package under the Virgin brand. Other European quad-play service launches include E-Plus/Kabel Deutschland in Germany and Telenet in Belgium.

In the United States, four of the largest cable companies—Time Warner, Comcast, Cox Communications and Advance/Newhouse Communications—have linked up with Sprint Nextel and will start offering wireless services directly to consumers in 2007.

Key Success Factors

Most of the initial quad-play announcements are emphasizing price discounts. While this seems like a natural first step, operators should also pay close attention to other business aspects to maximize chances of success.

Watch Your ARPU Carefully

Reduced churn should not come at the expense of insufficient average revenue per user (ARPU) or depressed margins. Therefore, service providers should adapt their quad-play discount strategy to competitive pressure and only offer heavy discounts on service bundles in markets where there is strong competition; such market conditions exist in the United Kingdom and France. On the other hand, there is less incentive to compete on price alone in less-competitive markets such as Belgium. In the United States, although competition between cable companies and telcos is becoming fierce, customers are used to paying significant monthly fees for TV, which allows room for much more flexible packaging, especially when mobile fees are included.

Don't Forget about Combinations of Double Play and Triple Play

Not all subscribers will be interested in full quad-play services. Initially, some of them may forgo the convenience of single billing and create their own bundles by combining services from different providers. Other people may only be interested in two or three of the available quad-play services. Successful operators will recognize these demand patterns and provide combinations of double- or triple-play bundles for different consumer segments. For example, a triple-play combination could be mobile, Internet, and IPTV.

Select a Solution with Pre-Integrated Components for FMC and IPTV

Quad-play solutions can require complex integration, which may lead to long deployment schedules and uncertain return on investment (ROI). To increase chances of success, service providers should look for pre-packaged solutions of IPTV with fixed-mobile voice and video communication applications. Key components that should be pre-integrated include IPTV middleware, set-top boxes (STBs), network personal video recorder (NPVR), billing, fixed-mobile telephony and video, instant messaging, presence, interactive video services, and media center connectivity.

Get the Fundamentals Right

Deploying quad play is a serious investment for any service provider; it is therefore important to pay particular attention to fundamentals such as the following:

- Content acquisition, management, and distribution
- High-quality operations and customer support
- FMC architecture with a future-proof evolution to IMS and true session initiation protocol (SIP) interoperability

Value-Added and Interactive Services

Overview

The real goal of quad play is to provide more than four services at a lower cost and on a single bill. Quad play gives operators the ability to create additional subscriber value by delivering value-added services seamlessly over any device, providing the potential to generate higher revenues and margins.

Examples
Seamless Video Services over Fixed and Mobile Networks
Today's mobile handsets are more than just telephones—they have also become wireless video camcorders and videophones. This means that video services, including video calling between mobile, TV, and personal computer (PC); "see what I see" from mobile to TV; and video mail from mobile to TV, can now be made available across all devices. French operator Orange has been a pioneer in the implementation of FMC video telephony and interactive video between third-generation (3G) mobiles, PCs, and videophones.

Dual-Mode Handsets
Dual-mode handsets are a key component of fixed-mobile services. They can be used to make voice or data calls on a cellular network in a mobile environment and over the Internet while at home. To optimize network connectivity, the handsets switch from cellular outside to a wireless fidelity (Wi-Fi) connection inside and also add the advantage of a single fixed-mobile address book. According to In-Stat, more than 132 million Wi-Fi/cellular handsets will be sold in the next four years. Service providers actively promoting dual-mode Wi-Fi/cellular handsets include Free, Neuf, Orange, and BT.

Messaging across Multiple Devices
Messaging services such as e-mail, short message service (SMS), and instant messaging (IM) have proven to be true killer applications. A natural FMC evolution is to offer transparent messaging between cellular phones and PCs. This approach lets subscribers enjoy the same interface, along with a similar look and feel, on all mobile and fixed devices. For example, in June 2006, the mobile virtual network operator (MVNO) Ten was launched in France with a focus on packaging voice minutes with mobile e-mail and IM (through MSN).

Interactivity between Wireless Handsets and IPTV
Mobile handsets will be increasingly used to preview TV content or recordings on home digital video recorders (DVRs) or remotely instruct DVRs to record programs. Sprint has indicated that these types of interactive services will be available to residential customers using Sprint's Power Vision handsets in 2007.

Key Success Factors
The value of the services listed above can be enhanced by focusing on the end-user experience and using sound business principles.

Use Personalization to Go beyond Low Pricing
Discounted pricing for quad-play service bundles offers little room for differentiation, so it is in the interest of operators to emphasize innovative, personalized applications.

One challenge with the converged quad-play model is the mobile component. The triple play of fixed telephony, broadband access, and TV services is a household purchase, but mobile is more of an individual purchase. The solution lies in the personalization of other service components such as IPTV and Web surfing. IPTV differs from traditional TV programming in a number of ways, including more viewer control over and involvement with the service. Personalized TV viewing, e-commerce, gaming, and electronic voting will become standard features of IPTV. Additional personalized services will be created. For example, Google recently announced that they had developed a way to use audio samples to quickly identify which programs a person is watching on TV to deliver personalized content to the TV or PC based on that information.

Focus on Innovative Services to Drive High Revenue and High Margins
The full potential of quad play is ultimately realized with innovative services that combine content and communication across all types of devices and all types of networks (IP, Wi-Fi, 2G, 3G, worldwide interoperability for microwave access [WiMAX]).

For example, interactive video represents an untapped potential of high-margin services across mobile/TV/PC, including dating, video blogging, televoting, home surveillance and monitoring services, and betting.

To identify the winning applications, service providers will need to pursue a trial-and-error strategy, testing different technologies, service combinations, and marketing bundles in different market segments. To maximize the success of these trials, it is critical to work with an ecosystem of service developers and partner with innovative suppliers.

While large operators are often in the best position to fight in a price war, it is interesting to note that smaller service providers have an edge when it comes to deploying new applications rapidly and trying out new service combinations and marketing approaches.

Lifestyle Services

Overview

In the past, the network was king and there was little choice for end users: a few operators provided the same services over the same telephones to everyone. There were specialist providers for communication, entertainment, and information services, and different services were available over different devices. Quad play, with personalized features and custom devices, enables services to evolve toward a world in which the user is king. Successful service providers will be the ones that are able to offer lifestyle service bundles targeted to groups of users according to their interests and usage patterns.

Quad play is the enabler of these new lifestyle services. By definition, quad play enables seamless access over all devices of personalized lifestyle services.

Examples

Personalized Lifestyle Services for Different Age Groups

The following service bundles are designed to capture a larger share of consumer spending by increasing the personal utility of communication and entertainment according to age segments:

- Eight- to 15-year-olds: instant messaging (IM), video greetings, gaming, multimedia ringback tones, SMS/MMS, friend locator
- Fifteen- to 25-year-olds: Video greetings, video blogging, video "see what I see," personalization of ring tones and phones, SMS/MMS, IM, e-mail, televoting, friend locator
- Twenty-five- to 55-year-olds: Parental controls, teen locator, single directory, personalization of

ring tones and phones, speech dialing, security, work productivity

Most operators have started offering subsets of personalized services, but the full potential is yet to be realized.

Lifestyle Services for Families with Young Children

These services include family-branded screen themes, ring tones, wallpapers, and games. Other key elements are child locator services and parental controls to manage family phone usage and control children's phone usage. Family lifestyle services require a well-established family brand. This is why Disney is the main contender with the recent launch of its Disney Mobile service. A natural evolution will probably create stronger links to other elements of the Disney brand such as the TV channel, radio station, and Web site, as well as broadband connectivity, finally making the MVNO/VNO model profitable.

Lifestyle Services for Seniors with Chronic Health Problems

The expanding elderly population has created a growing market for personalized health monitoring and alarms to offset the escalating costs of chronic care. Wireless devices can now transmit situation-based and location-based data in real time for online monitoring, analysis, and care. Orange was the first operator in the French market to launch a service to assist Alzheimer's disease patients. The service consists of a wristwatch-sized telephone and GPS monitoring device that sends out an alarm if the patient leaves a predefined geographic area.

Key Success Factors

Operators have to build a good understanding of specific consumer segments to build profitable business models for lifestyle services. To maximize potential, the following principles should be applied:

- Branding use in line with the targeted lifestyle market
- Custom user interfaces or custom-designed wireless devices
- Links with complementary lifestyle elements such as Web sites, IPTV channels, radio stations, original content, games, books, films, and merchandising

Conclusion

While it is difficult to tell in advance what the winning service combinations will be, having the potential for full quad-play services and a well-defined strategy will increase any service provider's chances of success.

Initial deployments have focused on price discounts. Soon these will be replaced by more advanced applications and service bundles, which will form the basis for mass customized lifestyle services for communication and entertainment.

With quad-play capabilities, both large and small players are in a stronger position to provide the right combinations of double-, triple-, and quad-play service bundles in line with consumer requirements.

The Triple Threat and the Quadruple Play

Randy Lay
Senior Vice President
Buccino & Associates, Inc.

It has never been easy to navigate a business through the ups and downs of the telecommunications marketplace, and it is not getting any easier. Developments in three key areas—regulation, industry consolidation, and the introduction of disruptive technologies—have created opportunities and new challenges for every participant in the industry. These developments, the "triple threat," also make it more challenging to offer sound advice and counsel to industry participants.

On the regulatory front, the Federal Communications Commission (FCC) has, with singular clarity, issued a series of decisions that have defined the competitive "level playing field" as one characterized by inter-modal competition, i.e., competition for communication services customers between the cable companies and the large, incumbent telcos, the so-called regional Bell operating companies (RBOCs). In its decisions, the FCC has come down clearly in favor of reducing the regulatory framework imposed on the incumbent firms. A key aspect of this framework was the requirement that the incumbents provide infrastructure to competitive firms essentially at cost, and the regulator views this as having caused the incumbents to reduce their investment in new infrastructure and technologies. The opposing view is that the FCC has in fact acted to restrain the competitive forces that were unleashed by the Telecommunications Act of 1996. In FCC Chairman Martin's remarks at the spring 2006 CompTel Convention and Expo, a competitive communications industry trade organization, he described this regulatory approach as

regulating "down" from the RBOCs rather than "up" from the competitive telecom industry—the small and not-so-small companies that sprung up after the 1996 Act. Of course, many of these competitive companies figured prominently in the telco meltdown that began in late 2000, so it is the survivors that are now impacted by this approach.

The practical impact of this regulatory climate is the increase in the level of uncertainty surrounding the business plans of mid-market competitors, particularly those that provide services by using infrastructure leased from the incumbent providers. In this climate of uncertainty, the challenge facing these companies is to maintain profitability, find new financing, and establish strategic alliances to provide additional services and replace leased capacity.

Not surprisingly, this has led to the second key development in this marketplace, a wave of consolidations that has seen the RBOCs acquire the competitive long-distance/diversified communications players, i.e., Verizon/MCI and SBC/ATT/BellSouth, while the competitive communication space has seen consolidation driven by Level 3, which has used a combination of debt and equity to acquire several competitive local-exchange carriers (CLECs) and lit fiber-optic providers.

The third key development is an explosion of technological innovation in communications, a field of play where the financial markets, rather than regulation, are calling the likely winners and losers.

These technologies—worldwide interoperability for microwave access (WiMAX), voice over Internet protocol (VoIP), IP television (IPTV), and fiber-to-the-x (FTTx) have all been recently in the news regarding developments at SprintNextel, Vonage, AT&T, and Verizon. All require significant investment to deploy, and, as the Vonage initial public offering (IPO) has shown, the public markets can approach the business plans for the commercialization of these technologies with a significant degree of skepticism.

The communications industry has never been a place for the meek, and the latest example of this is the quadruple play, a services bundle offered by the cable companies that includes high-speed data, plain old telephone service (POTS), TV, and wireless telephone service. This is a way to take advantage of the cable companies' installed infrastructure, drive out other competitors from the household (dead aim at the RBOCs and other telecom incumbents), and reach the communications nirvana of low customer turnover in a monthly recurring revenue business. It is also a very difficult offering for a mid-market communications competitor to match if they do not possess the combination of installed infrastructure, access to content, and technology.

So, how can the mid-market competitive provider be successful? And how can its financial and strategic advisors bring their experience, contacts, and creativity to bear and help them to be successful?

First, take an inventory of the basics, which are commonplace but not necessarily common. Firms that grow and generate more cash than they consume succeed; those that do not, fail. Firms that sell products and services that attract a broad base of customers do better than those with significant customer concentration. Savvy management teams that have a clear view of strategy and an ability to execute manage companies that have a better chance for success. Performance on these parameters is the foundation for success, either as an independent or as an acquisition candidate.

Second, in this industry and at this time, make the hard calls. Does your company have a business that can stand on its own, or does the business have products or intellectual property that would be better developed and exploited as part of another entity? If so, drive the business to maximize value and find a home.

Finally, action always beats inaction. For a whole host of reasons, not the least of which is the challenging state of many a balance sheet and profit and loss (P&L) in this industry, completing the assessment of where the firm is on financial and strategic grounds and acting accordingly will be time well spent. Swift and sure action positions the competitive mid-market firm to take advantage of opportunities in a rapidly changing environment.

Redefining the Quad Play with IPTV and IMS

Matthew Marnik

Marketing Director, Broadband Services and Multiplay
Juniper Networks

Introduction

As service providers add interactive video services to their voice, data, and wireless triple-play bundles, will their subscribers be watching Internet protocol television (IPTV) or engaging in a multimedia experience?

There is an elegant and cost-effective way to not just deliver a voice, data, video, and wireless bundle, but also seamlessly integrate them into a richly interactive and personalized experience.

At one time, a service provider could differentiate itself by offering the triple play: three services—wireline, wireless, and Internet access—bundled on one bill. However, the triple play is becoming commoditized. Further, service providers are stepping up to deliver the ultimate quadruple-play bundle by adding IPTV video services.

The quad play bundles wireline, wireless, Internet access, and video in one service package with one monthly bill. The goals of offering a quad-play bundle are as follows:

- Win a greater share of wallet—Monthly revenue per subscriber can increase more than 85 percent when a telco adds video services to its voice and data offerings, according to Yankee Group research (Will Video Drive New Revenue Growth for Telcos?" Yankee Group, May 2004).

"Households that subscribe to premium channels, digital cable and phone, and Internet services are more likely to stay loyal [and] can easily rack up a bill of over $120 a month," Peter Grant reported in the Wall Street Journal ("Cable Trouble: Subscriber Growth Stalls as Satellite TV Soars," August 4, 2004).

- Reduce churn—"Adding a strategic product (voice, data, or video) can reduce churn by approximately 25 percent or more," say Patrick Mahoney and Kate Griffin of the Yankee Group (Driving toward the Triple Play: The Telco Video Opportunity, Consumer Technologies and Services, Yankee Group, September 2004).
According to the report, bundling digital subscriber line (DSL) and long-distance service reduced churn by more than 50 percent for Bell South and 70 percent for SBC. Adding a cable modem product reduced cable TV churn by more than 25 percent for Cox Communications. When Cox completed its triple play by adding voice, churn dropped by more than 50 percent to an enviable 0.7 percent.

The Problem with the Traditional Business Model
Both cable and wireline operators are pursuing quad-play offers to maintain a competitive edge. Cable operators are enhancing their video on demand (VoD) offers and pursuing partnerships to add wireless to their existing video service and emerging voice services. Local telephone compa-

nies are adding high-definition television service to their established voice and Internet services. Both types of operators are hoping to win by offering customers the convenience of one vendor and one monthly bill at an attractive price, compared to buying each service individually.

The trouble is that the services are still separate everywhere except on the bill in the customer's mailbox. In the traditional business model, the reality is as follows:

- The networks are separate—Wireline access has traditionally been about voice. Broadband access has been a data service that is moving toward entertainment. Wireless started with voice but now delivers data and entertainment. IPTV is about video but is moving to add voice and data. A provider may be maintaining two or even three network infrastructures to deliver all of these services.
- The user experiences are separate—Content and devices are tied to access. Subscribers use different devices, interfaces, and methods to access their various services, which are different and do not interact with each other. The learning curve is steep for many users to master services on a new device. Users are demanding services that are easier to adopt.
- The billing systems are separate—Subscribers may see all their service charges integrated on one bill, but behind the scenes are disparate billing systems that must be maintained separately for each service.

With all this duplication of infrastructure, operating expenses are high. It is complex and costly to introduce new services. Even if you could create a consistent user experience for a service across multiple media, the application would have to be developed separately on each platform. For users, there is little benefit in moving all of their services to one provider except for the promise of a better price.

Worse yet, this quad-play offer is easy to duplicate. A competitor can come along and bundle exactly the same services at a lower price. If you are a service provider, where is your edge?

Compete with an Enhanced Service Experience instead of Price

If you are bundling discrete services and hoping to win customers on the allure of bundling, you are preparing for a price competition. How long do you want to "win" by eroding your own profit margin, especially if the company is footing the costs of separate infrastructures for these bundled services?

With advances in Internet protocol (IP) and IP multimedia subsystem (IMS) technologies, there are now better options, including the following:

- IP provides a cost-effective way to converge voice, data, and video transport onto a unified network infrastructure.
- IMS provides the next-generation core architecture that converges voice, data, and IPTV service attributes over multiple access types into one consistent user experience that is independent of a user's access or device.
- IMS takes the quad-play offer to the next level. By linking IPTV with IMS, television set-top boxes (STBs) can be added to the list of IMS endpoints, along with mobile phones, personal computers (PCs), and other consumer entertainment devices.
- Users can enjoy a consistent user experience across various devices and access networks. Voice or data services can be extended to IPTV with the same look and feel as on a session initiation protocol (SIP)–based wireline or wireless device. Providers can offer services that help users manage their personal libraries of commercial and private content and extend these services to multiple devices and access networks. Video services such as network-based digital video recorders (DVRs) can be extended from the TV to mobile devices, with the goal of enabling users to take their content wherever they go.

Now the service provider is not competing solely on price, but also with an integrated offering that offers a richer user experience. By increasing the value of its service offer and not just the service inventory, the provider can gain significantly higher average revenue per user (ARPU) and loyalty.

Will Consumers See Value in an Integrated Quad-Play Offer?

Let us look at that last application—mobile TV. What if a provider in your area offered a service that allowed you to keep all your current TV channels but also watch them on your cell phone, PC, laptop, or car device? Would you sign up?

When Harris Interactive Research asked this question in a recent survey, 15 percent of respondents said yes and 9 percent indicated that they were willing to pay a premium for the privilege. Students and young singles were particularly intrigued, as were households with younger children, especially if they already had broadband service.

Mobile TV is just the beginning. A service provider that offers IPTV with an IMS foundation has many more opportunities to differentiate itself with specialized services. All of these services can be personalized by the user, including mobile gaming, music subscriptions, and video chat overlaid on live content.

New Service Opportunities with IMS–Enhanced IPTV

The initial opportunities for service convergence include adding voice and data functionality to IPTV and extending IPTV functionality to mobile devices. Beyond this, you can extend core service enablers to create a unified user experience across multiple devices. The following subsections describe typical opportunities that are possible with the combination of currently available IP and IMS technologies and an STB or card for the user's television.

Extending Voice Services into the TV Environment

While watching TV, the user receives an on-line prompt showing the caller ID of an incoming call. The phone does not ring (a user-selectable setting) so that nobody else in the household is disturbed. The subscriber can choose from the following options:

- Accept the call. The phone rings and the call can be answered on the home phone, a mobile phone, or a speakerphone associated with the TV.

- Reject the call. The phone never rings and the call is discarded.
- Forward to voice mail. The caller can leave a voice mail, and the subscriber sees a message-waiting indicator on the TV screen.

The subscriber does not have to get off the couch to see who is calling, which is particularly welcome if the calling party is not of interest. Who wants to interrupt a favorite TV show to take a message for another household member when voice mail can do just as well or better?

The service also supports click-to-call capability, whereby a subscriber can place a call using the remote control, either from an address book or a list of received calls. In the future, subscribers will also be able to place and receive video calls using a TV-mounted Webcam.

Extending Data Services into the TV Environment

Data integration can also take several forms. For example, subscribers could exchange instant messages (IMs) with others on their personal buddy lists while watching a TV show. Thanks to "presence" capabilities, the system knows whether the subscriber's buddy is watching the IPTV service, and if so, sets up the IM connection for them to share a back-and-forth chat while watching TV. The two subscribers can then share the viewing experience and trade comments about what they are watching, even though they are in different places.

Picture sharing is another popular option. Suppose you want to share digital photos with a distant friend or colleague who does not have a PC or Internet connection. You could send the photos to the service provider's photo exchange service, which in turn uploads them to the recipient's STB. That person receives a notification that pictures are available for viewing, and selects "slideshow" on the TV remote control to view the photographs on the TV screen.

Extending TV Services into the Wireless Environment

A mobile device can become an extension of the IPTV service—both to control and to view video content. For example, suppose a change in your flight schedule will cause you to miss tonight's episode of your favorite TV show. You can use your PC, personal digital assistant (PDA), cell phone, or other wireless device to pull up the TV schedule, select the episode, and send a command to the DVR service to record this episode.

Later, you can use your video-enabled mobile device to call up the DVR menu, select the pre-recorded program, and watch it wherever you are. Conversely, you could begin watching the TV show at home and then switch seamlessly to the mobile video device to watch the rest of the program on your trip. The user interface will be the same on the mobile device as on the home IPTV set, making navigation familiar and convenient.

Establishing a Consistent User Experience across Access Media and Devices

The unified service experience is made possible by core service enablers such as the following:

- Presence—Subscribers can see if someone on their buddy list is on-line, whether that person is connected through a mobile device, a PC, or an IPTV.
- Network buddy list—This is the subscriber's circle of cohorts for sharing interactive services. The buddy list would be the same on a PC, a mobile device, or an IPTV.
- Single sign-on—Users can log on to a service using one device and continue their sessions in another device without having to sign in again. For example, when a subscriber signs on and accesses IPTV from the TV, there is no need to sign in again when switching over to watch the rest of the show on a mobile device. The service is seamless among devices.

A Typical Example of IMS–Enhanced IPTV Services in Use

The following is a hypothetical example of how these integrated capabilities can be used in the real world to transform the entertainment and communication experience.

Mary, who is in the middle of a big remodeling project, is watching a home improvement show for inspiration. Her personal avatar appears on the television screen and plays a short multimedia video clip of her interior designer, who is also watching the show and wants to share some ideas based on the home they are viewing. Mary uses the remote control on the STB to accept the incoming session. A picture-in-picture window opens up, and Mary sees her designer appear.

Together, they agree on some light fixtures and decide to open a group chat session with Mary's husband, Greg, who is at the home improvement store. They text-chat with Greg to ask him to check out fixtures of the type they just saw on TV. Using his video-streaming mobile handset, Greg captures some short video of the lighting fixtures, and the IMS application plays the video clip to both women simultaneously.

Recognizing the presence of an active group session, an advertisement application linked to the home improvement show sends out an overlay video clip offering the chance for the group to sign up, for a small fee, to participate in an information session about selecting and installing home lighting. Greg and Mary accept and enjoy the multimedia learning session from their mobile handsets and TV sets.

Mary then uses her handset to instruct the IPTV application to track remodeling ideas on her subscribed channels. While Greg and Mary are at work, the network-based recording capability creates an index of the targeted content. Whenever they choose, they can readily navigate through the recorded program choices and access them from any of their IMS devices.

When offered with an IMS foundation, IPTV changes the session from simply watching TV to experiencing TV—from a passive and solitary activity to a richly interactive one.

Benefits of the IMS–Based Quad Play

For service providers, IMS–enhanced IPTV offers a number of benefits, including the following:

- Winning new subscribers (and keeping the ones they have) by offering a richer home entertainment experience than can be achieved with competitors' standard triple- or quad-play offerings.

- Gaining new revenues by delivering a differentiated, value-added service—personalized and interactive—supplemented with revenues from carefully targeted advertisements.
- Reducing operating expenses by reusing a variety of functions across the quad-play environment, such as subscriber, service, and user profile data; authentication; authorization; digital rights management (DRM); charging support; and the media and data servers that optimize delivery to various device types.
- Deploying new services faster, since an application can be created in one place and deployed across all access networks and devices, with a core billing system that aggregates billing data.

For end users, IMS–enhanced IPTV improves the communication and entertainment experience with value-added capabilities. The traditional TV viewing experience can be combined with diverse types of person-to-person or group communications such as chat, instant messaging, caller ID, videoconferencing, or video mailbox to enrich the experience.

Subscribers can create, manage, and share their unique libraries of content—both commercial and personal. For instance, they can establish video surveillance for home security, share photographs and video blogs with friends, push Web pages, or send a favorite, prerecorded TV show to a fellow subscriber.

In spite of the diversity of service opportunities, the user experience is simpler—and familiar from one device to another—and services can be personalized to meet users' needs across devices.

The Architecture of the Converged Quad-Play Environment

How can an IMS core network be a key enabler of a converged IPTV/quad-play service offer? The following is a high-level view of the IMS core elements in delivering what we call the "converged quad play":

- Single subscriber authentication—This enables a user to log on to a service using one device and continue the session on another device without having to sign in again. This capability

is made possible by a master database managed through a home subscriber server (HSS) element in the IMS core network.

- Unified session control—This enables sessions to be handed off between devices, such as when an IPTV viewing session is started on a TV and then continued on a mobile device. This function is accomplished from a call session controller (CSC) element in the IMS core network.
- Service enablers—Including presence and network buddy lists, these can be extended to any network device via the IMS core network, enabling features to be delivered in the same manner to a PC, mobile device, or IPTV STB.
- Application ubiquity—This refers to the way the IMS application layer applies to multiple access networks. New applications and services can be extended from one access or device type to another with relative ease.
- Resource control—This ensures that the required resources for a service are established among different access types. This IMS element manages policies for service by access type and makes sure the specific access network can deliver the required attributes.

IMS provides a core network foundation that allows a service provider to deliver a personalized multimedia experience beyond basic IPTV. Prior to IMS, a typical deployment used proprietary interfaces, separate service and subscriber processors and databases by access type, and separate networks for voice and data.

With IMS, the in-home television (STB) becomes another class of IMS–enabled device, much like mobile phones, PDAs, and laptop computers. Service providers can extend advanced, multimedia IPTV services to IMS–enabled devices—any content on any device—converging voice, data, and video services into a differentiated service offer.

IMS open-standard solutions use a single source for authentication, quality of service, content management, and security for telephony, multimedia, and IPTV. It enables users to sign on only once to access services across multiple access types, and it supports personalized libraries of content to be delivered to the subscriber's TV, PC, PDA, or other video-enabled mobile device.

Conclusion

It is not enough to create a basic quad-play service offering that delivers four separate services on a single bill. Yes, customers do want the convenience and simplicity of a single provider—and they do appreciate the discounts they can get with bundled services. If you are willing to compete solely on price, this is the way to go.

However, customers are eager for more than a price break and a single bill. They are looking for simplicity in managing and using their various services and devices—and they are willing to pay a bit of a premium for it. IMS can get you into this emerging market quickly.

Future-Safe Networks – The DNA² Uncertainty

Haim Melamed

Director, Channel Marketing
AudioCodes

Triple play, fixed-mobile convergence (FMC), and Internet protocol multimedia subsystem (IMS) are the three main drivers of the telecommunications world today. Quadruple play, which is actually triple play and FMC bundled together, is what carriers are moving toward, providing multiple, optionally bundled, services to customers. The third—IMS—is the unified future network, based on IP, required to seamlessly support various services of quadruple play or any subset of quadruple play.

Any telecom expert that you encounter will agree that voice, video, and data in the fixed and mobile environments will eventually ride over IP. Convergence is apparent, allowing additional services to run on a single IP network.

Are we close to a point where one solution will fit all? Surprisingly, the answer is no. Any carrier that is currently building a next-generation network (NGN) is facing multidimensional uncertainty.

The network devices, network, access, and applications (DNA²) future is actually diversifying rather than consolidating. Stronger and more feature-rich handheld devices and customer-premises equipment (CPE) products have been created. New network-based protocols and codecs emerge each day. Different types of broadband access technologies exist (e.g., digital subscriber line [DSL], cable, fiber-to-the-x [FTTx], broadband over power line [BPL], broadband wireless access [BWA]). Many new applications and services need to be deployed rapidly to increase average revenue per user (ARPU) and reduce churn.

Any carrier that is trying to build a "future-safe network" under this multidimensional uncertainty is facing a great challenge. Selecting a single vendor as an infrastructure provider will possibly "lock" the carrier to specific devices, network, access, and service creation environments. Moving toward the "best of breed" direction, as derived from the IMS architecture, has proven to be the correct course forward.

Devices

Comparing the telecom world to the information technology (IT) world exhibits similar behaviors over time, with the telecom world lagging behind the IT world. Looking back at history, the IT world has moved forward from a dumb terminal-based world, connected via a fixed connection over a slow network to a centralized mainframe computer. This new world encompasses high-performance personal computers (PCs) communicating peer-to-peer via a wireless, fast network.

Similarly, the telecom world has moved from the old plain old telephone service (POTS), connected via a fixed-line, voice-only time division multiplex (TDM) connection, to a world where mobile devices supporting voice, data, and video communicate with each other over high-speed, third generation (3G) networks.

Current telecom devices are hybrid devices, supporting multiple applications in a compact form. The central processing unit (CPU) power in today's mobile telecom devices equals the power we had five years ago in home PCs. In parallel, the speed of today's cellular 3G networks equal the speed of local-area PC networks that existed 10 years ago. Today there is storage space in mobile devices that is equivalent to storage we had in home PCs five years ago.

The combination of powerful edge devices on one hand, and a very fast network on the other hand, enables the creation of many applications. Distributed computing, file sharing, videoconferencing and streaming, and additional applications are available today in our home PC over broadband Internet connection, are becoming available on the mobile telecom devices too. The life cycle of telecom devices today is between nine and 30 months, depending on the geographical and cultural elements. However the replacement of the user's mobile device will create new revenue for the service provider, while the substitution of a network infrastructure supporting new device applications leads to expenses being incurred. Therefore, service providers must become vigilant while building an infrastructure correctly, as this will support robust, faster, and application-rich devices.

Network

Building a network to support new applications and devices is not an easy task. Although some levels of uncertainty have been eliminated, we are left with many unanswered questions.

IP is the chosen protocol for all future networks. Naturally, voice is carried on these networks using voice over IP (VoIP) protocols. The architecture of IMS as the core solution for NGNs enables the interoperability and integration of different network elements between themselves and between various network services and applications.

The interaction between new-world devices and the NGN IMS network creates many new media types, formats, and protocols. On top of the core IP–based network we see multiple control protocols (e.g., session initiation protocol [SIP] and H.248), multi-

ple media types (e.g., voice, video, data, audio, images, animations), and numerous media formats: G.711, G.722, G.723, G.726, G.728, G.729, G.729e, enhanced full rate (EFR), adaptive multirate (AMR), wideband AMR (WB–AMR), enhanced variable-rate codec (EVRC), QUALCOMM code excited linear predictive (QCELP), fourth-generation vocoder (4GV), Internet low-bitrate codec (iLBC), image save and carry (ISAC), H.263, H.264, Moving Pictures Experts Group 4 (MPEG4), and more to come.

Building a network to allow the transparent move of all of these and many others in the future involves a lot of open questions. A service provider must select the right equipment for its IMS infrastructure to be able to support all of these and many unknown others in the future without a forklift upgrade.

Access

The advantage (and disadvantage) of IMS is that it does not define the access technology. The greatest variety in the telecom world today is of access technologies. In the wireline space, other than the TDM telephony network, the dominating technology is broadband. Among the different broadband technologies we can find dozens of variations of DSL, cable, fiber, and BPL. In the wireless space, we can find many variations of cellular broadband wireless technologies, wireless LAN technologies, and fixed-wireless access technologies. A partial list of these technologies includes Universal Mobile Telecommunications System (UMTS), code division multiple access 2000 (CDMA2000), 3.5G, 4G, TTD, wireless fidelity (Wi-Fi), MetroFi, worldwide interoperability for microwave access (WiMAX) (2004/16e), Flash–OFDM, and many future additions.

Although all of the above-mentioned technologies are the transport technologies for voice, video, and data packets over IP, each one of them requires the infrastructure to support different network characteristics (e.g., bandwidth, delay, jitter, packet loss) and different media codecs.

In addition, the use of all of these access technologies and protocols in parallel require massive resources of transcoding and protocol translation in the access and in the network. A solidly built, future-safe NGN must take into account these requirements and the uncertainty of the future winning access technologies.

Applications

Customer applications are the actual drivers of the multidimensional uncertainty of NGNs. Increased ARPU and reduced churn are the main concerns of telecom operators, and the only way to achieve this is to introduce new applications.

Telecom applications come in waves. Globally, we can see different waves of applications occurring at different times.

The first wave starts from voice, followed by text, data, video, and multimedia. Our parents used the telecom infrastructure for voice calls only. We are using it today extensively for text messaging too (mainly short message service [SMS]). Our children are now using it to browse the Internet and download music and games, while some of us are already using it for video calls and video streaming.

The second wave begins from real-time use of the network and adds the support for store and forward applications. For each type of telecom media, we have the real time and the store and forward service versions: a phone call versus a voice mail, a chat versus an SMS or an e-mail, a video call versus a video mail, and live video broadcasting versus video on demand (VoD).

The third wave starts from 1:1 applications and adds the support of group applications. Once again, all types of media have both versions: a phone call versus a conference call, a personal chat versus a group chat, a personal SMS versus a group SMS, and a video call versus a videoconference call.

All of the above services must be supported on the future NGN concurrently. Each of them require different device capabilities, different network services and protocols, and have different characteristics on top of different access networks. Again, a service provider planning his NGN must take into

account the evolving new applications and plan his NGN to support all of those in the future.

End-to-End versus Best-of-Breed

Understanding the multidimensional uncertainty in building an NGN, a service provider can take one of two approaches: the easier approach is to work with one solution provider for most (if not all) of his network components. The other approach is to adopt the best-of-breed architecture and build an open, standard-based network to allow future flexibility in devices, network, access, and applications.

In the short term, working with a single solution provider is an easier and safer approach—no integration hassle, no interoperability problems, and short time to market. On the other hand, the essence of the NGN architecture is the openness and the ability to introduce future, unknown applications and technologies quickly and easily. Working with a single vendor will possibly lock the service provider into some proprietary architectures and protocols. It will make it harder—and in some cases impossible—to integrate other vendors' solutions into the network, and most important, it will not leave the freedom of choice in the hands of the service provider.

Choosing the best-of-breed approach is not an easy choice. It does involve more integration efforts in the short term and introduces the overhead of system integration challenges. On the other hand, implementing a standards-based, best-of-breed NGN will allow the choice of the best available network elements, services, and applications, and the future introduction of any device, access network, or application supporting standard protocols and services. IP, VoIP, and IMS make this option the easiest ever.

Combining these two approaches yields an interesting situation whereby a service provider makes a hard requirement from its solution vendor to become an integrator and use best-of-breed elements for important parts of the solution network elements such as application servers, media gateways, softswitches, and media servers.

The Importance of Minimal Best-of-Breed IMS Implementation – the Media Gateway and Media Server

In the world of best-of-breed IMS telecom networks, two of the most important components of the carrier networks are the media gateways and media servers. These devices are actually the only devices in the network that mediate between the different access protocols and technologies, between the different media types and codecs, and connect between the TDM and the IP world. The media server function is the hardware resource for many NGN applications.

Selecting the right media gateway and media server for an NGN today captures every aspect of the multidimensional uncertainty. A future-safe media gateway and media server must support existing and future control protocols, media types, codecs, and PSTN protocols to be able to support the evolving NGN technologies into the future.

UETS: The Multiple-Play Next-Generation Network

José Morales Barroso
Director
L&M Data Communications

It is quite difficult to break with pre-established ideas, even more so if people have followed the "doctrines" in use for a long time. But to progress, another point of view, another perspective, is needed; this means forgetting previous conditioning and prejudice in building for the future, to take advantage of the once-in-a-lifetime opportunity offered by the next-generation network (NGN). Nowadays, everyone agrees that the future of networks is all–Internet protocol (IP). However, we should ask ourselves whether the future might be all-Ethernet. The Universal Ethernet Telecommunications Service (UETS) follows that philosophy. It is based on a fundamental idea that allows the convergence of all the digital services over a single Ethernet-based infrastructure, which uses an extraordinarily efficient and easy system to put it in practice. Thanks to a new switch concept described in this paper, a new world of opportunities opens up, making it possible to create a network supporting 70,368,744,177,664 interfaces on a single domain, at the same time providing notable improvements in speed, security, and scalability.

In the proposed system, network nodes will switch Ethernet frames directly at the physical layer, using the local media access control (MAC) addresses [1] and making it unnecessary to use routing, bridging, or signaling. This means packet-mode transmission with circuit characteristics, which provides several notable improvements to the state of the art. In other words, we are dealing with the development of a multiple-play universal service, combining the characteristics of the Internet, the public telephone network, virtual circuit networks, local area networks (LANs), voice-over packets, power line communications, wireless and mobile systems, and cable networks. This development will offer integrated services using existing infrastructures and create new ones as needed, thus making it possible to extend the Ethernet services to a planetary scale.

UETS has been presented at highly respected conferences such as the Terabits Challenges Conference at INFOCOM 2006, the ISOC Monthly Meetings, and the Global Internet Congress or eChallenges, a conference supported by the European Commission, and brings together delegates from leading commercial, government, and research organizations around the world to bridge the digital divide. The list of institutions and companies interested in this technology is growing.

Antecedents

Since 1970, a solution for what the International Telecommunications Union (ITU) named the integrated services digital network (ISDN) has been sought. The works produced in search of this solution resulted in the ISDN Recommendations; however, these did not achieve the expected results because the network was based on low-speed circuit technology (64 kbps), which is not suitable for data traffic. On the other hand, new systems are appearing that offer network access based on Ethernet/802.3 technology, but only as a service to connect access points. Illustrative of these are connection points to the network, such as in the Metro

Ethernet Forum. In this context, the 802.3 IEEE committee for Ethernet standardization created the 802.3ah working group [2]. This standard, approved in June 2004, is of use for the development of the system described in this paper, since it provides the specifications to establish connections using the public telephone twisted pair, a key element for reuse of the existing infrastructure. With 1.2 billion telephone lines in the world, there is a big potential for an integrated services solution that would use the traditional telephone's subscriber loop infrastructure and a voice-grade twisted pair at the user's location. For those instances where optical fiber is installed or needs to be installed, the 802.3ah EFM standard offers the possibility to establish connections over fiber optics ranging from 100 Mbps up to 1 Gbps, and 10 Gbps in the future.

Ethernet technologies are playing a key role in telecommunications, reducing costs, simplifying the user experience, and enabling new services for enterprises and consumers.

Where Does UETS Come From?

UETS originates from the comments that the author sent during his participation in the final balloting voting phase of the IEEE 802.3ah project, which were favorably received since "the suggestion may make a great new project" [3]. The UETS system came into being as a result of a further study and deeper reflection.

One of the fundamental bases of the invention is that—although Internet applications at the moment use the transmission control protocol/Internet protocol (TCP/IP) protocol stack, that corresponds to Layers 3 (L3) and 4 of the architecture of reference—the state of the technique makes it possible that these operate directly on the layer of connection defined in the standard IEEE 802, a solution to the unified network that offers UETS. It can be seen in a schematic way in *Figure 1*.

To develop UETS, the author has taken advantage of the accumulated experience of more mature technologies: the classic telephone network based on physical circuits, virtual circuit networks (X.25, frame relay and asynchronous transfer mode [ATM]), local-area networks (LANs) based on Ethernet, and TCP/IP. Choosing the best from every technology and combining it in a single new mode, according to the patented specifications, the author finally reached the solution, which has the following characteristics:

*   It is simple to put into practice, since it uses a single network technology, Ethernet, and at the

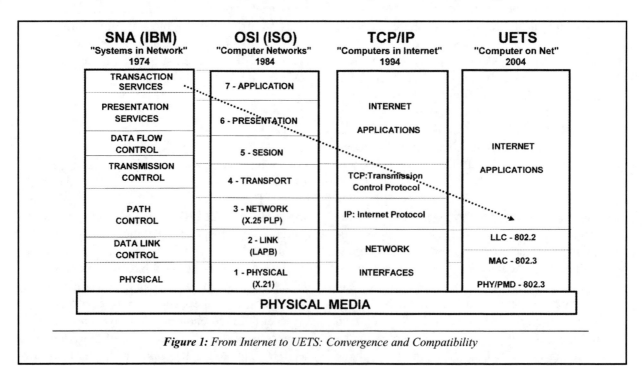

Figure 1: From Internet to UETS: Convergence and Compatibility

same time efficient in its functioning due to minimum loss caused by protocol overhead.

- It is able to support very high speed with low latency, using the central UETS switch.
- It is effective in flow and congestion control, prioritization, and quality of service (QoS), all integrated into the terminals and network nodes, guaranteeing real-time services.
- It is able to support current applications that run over TCP/IP without any modification.
- It is environmentally friendly, because it uses only the strictly necessary energy through its power management mechanism.
- It is able to guarantee service using well-tested devices and technologies, with a lot of practical experience, making it possible to develop the system with a minimum investment and risk-free deployment.

To sum up, the network proposed offers all the advanced communication services: voice, telephone service, data, videoconferencing, real-time video, video on demand (VoD), remote control, remote sensing, network storage, application server access, transactional services, network games, e-learning, e-medicine, e-commerce, etc., all of them using Ethernet technology, the media used by most of the systems and terminals deployed.

System Description

The main elements, represented in *Figure 2*, and characteristics of the UETS system will be resumed in the following points:

- It extends the Ethernet domain concept of the local scope, defined by the IEEE 802, to the service providers' network.
- A new access device, the terminator universal Ethernet (TRUE), is basically designed as a telephone but works in packet mode and includes typical telephone services such as emergency calls (necessary and very important for the telecommunications company) and power management for power saving. It also allows data terminal access to the network.
- A new terminal concept, the terminal universal Ethernet (TUE), runs applications using only a simple supervisor module and the graphical user interface (GUI) of the Internet navigator.

- With a new network node concept, the central office central universal Ethernet (CUE), these network nodes will be in charge of the service connection and the Ethernet frame routing at the physical level—based only on the MAC address—not requiring bridging, routing, or signaling protocols [4]. The possibility of handling several levels of priority, as well as the network flow and congestion control, is also contemplated by means of the combined operation of the CUE and the TUE. In this simple but indirect way, the dangers of the Internet just disappear in the Ethernet domain. No user can attack or impersonate another user.
- It can support any physical media that supports the transport of 802.3 frames and is capable of reusing all the present telecommunication infrastructures on campuses, in buildings, facilities, and homes, depending on the service requirements or the bandwidth needed in each case.
- It has the capability to integrate the entire communications infrastructure, including the voice telephone network, cable operator networks (coaxial), power line communication (PLC), fiber optics, and wireless and mobile phone networks.
- It is capable of remote powering via the twisted pair from the central office, making the emergency call service possible by means of batteries in the central office (this is the classic solution for the telephone network).
- The incorporation of a power management system for energy saving makes it possible to reduce the power usage to the minimum necessary for the UETS network, making a rational use of global energy resources.

Extension of the Ethernet Domain

A LAN is defined as "a computer network at the user's location, inside a constrained geographical area" [5]. The 802 LAN is defined as a "local-area network consisting of an access domain using the MAC protocol specified on the IEEE 802.n and ISO/IEC 8802–n standards." This applies to the Institute of Electrical and Electronics Engineers (IEEE) 802.3 and International Organization for Standardization (ISO)/International Electrotechnical Commission (IEC) 8802–3 standards, where the term Ethernet domain is taken from.

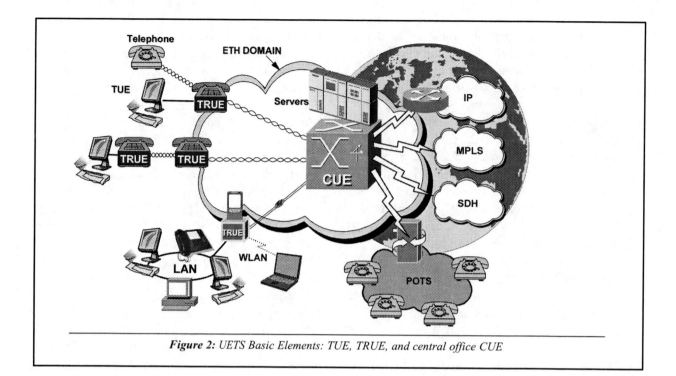

Figure 2: UETS Basic Elements: TUE, TRUE, and central office CUE

LANs based on Ethernet are used throughout the world. It is the primary data network; until very recently, it was restricted to installations on campuses or in buildings. A key point for the UETS is to extend the local Ethernet network to include the public operator's network. In this way, it can provide services similar to the ones available on LANs, as shown in *Figure 3*.

Figure 3a: "Classic" LAN with Ethernet Domain in Building/Campus

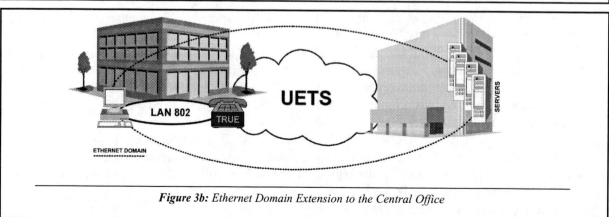

Figure 3b: Ethernet Domain Extension to the Central Office

In UETS there is a clear distinction between the Ethernet domain, which offers the service, and the IP domain, which offers Internet connectivity.

The group of network elements that uses MAC/802.3 to connect network users and servers is the Ethernet domain, limited to the service provider infrastructure and isolated from other environments, especially from the IP domain of the Internet. With this solution, the providers can offer services on the Ethernet domain, the IP domain, or a combination of both, taking advantage of all the current equipment and Internet environment applications and complementing the services offered exclusively by the Ethernet domain. This opens up new business opportunities to compensate for the decline of conventional telephony.

The Universal Ethernet Network Terminator (TRUE)

The network is accessed by means of the universal Ethernet network terminator, or TRUE, a device the user can connect to the twisted-pair connection of the subscriber loop, which is powered from the central office, or from whatever other physical means whenever it is necessary. The principal elements of the TRUE device are quite similar to those of an Ethernet telephone. In order to provide the basic telephony service, there is a handset that employs the voice-over-packet (VoP) technique. Access to the telephone network service can be offered and managed by the same operator, thus making it possible to maintain the same telephone number for incoming as well as outgoing calls.

In addition to the basic telephony service described above, users can connect terminals of diverse types directly to the TRUE device or through a LAN; thus, terminals can have access to the UETS network in order to use any of its services.

To be considered a true telecom service, the connection complies with the conditions of the classic telephone network: it guarantees the emergency call service (112 in Europe or 911 in the United States), it includes the necessary procedures for the maintenance and management of the link (operations, administration, and management [OAM]), and the terminal's power management, so that it only consumes power as needed, in contrast to the current Ethernet terminals, which consume power constantly. Powering the terminals directly from the central office is also more efficient than local powering from a power management perspective (see *Figure 4*). In addition, it is more reliable because it incorporates a higher level of redundancy and guarantees

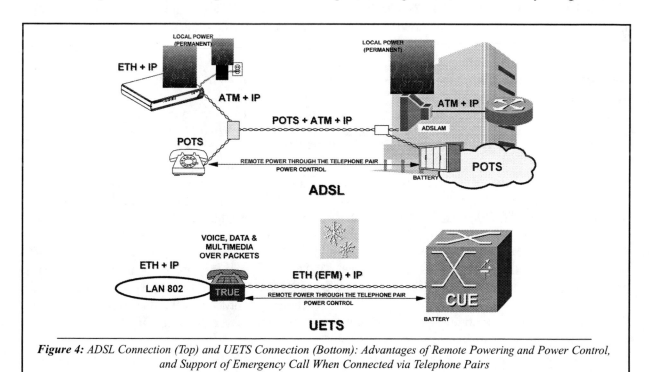

Figure 4: ADSL Connection (Top) and UETS Connection (Bottom): Advantages of Remote Powering and Power Control, and Support of Emergency Call When Connected via Telephone Pairs

the service with the batteries at the central office, the classic solution of the telephone network.

An Example of Energy Savings
If the current fixed telephone lines worldwide (<> 1.25 billion) are replaced by current broadband connections (<> 30 watts), with an average utilization of 5 percent (95 percent idle), there is a very high potential for savings of the energy control. Applying the proposed system, the savings obtained using the power management system would be as follows:

$$1.25 \text{ B x } 30 \text{ W x } 0.95 \text{ x } 24 \text{ h x } 365 \text{ d}$$
$$= 312 \text{ TWh/year}$$

The introduction of power control in the broadband access networks worldwide would mean an instantaneous savings equivalent to the production of 40 nuclear reactors.

The Universal Ethernet Terminal (TUE)
It is not a new idea to use a terminal to access network servers—it is actually a return to the beginning. What is really new and revolutionizing is the communications architecture of the terminal. By combining the TRUE and the UETS central office, it is possible to offer an extraordinarily simple and efficient service that is not dependent on geographical location and has an effective network congestion control.

The proposed terminal has the mechanisms to identify the network access point to the Ethernet domain, the terminal using the MAC address from the manufacturer, and the user employing a smart card (the user is also certified).

This way, network services are accessible from any terminal at any place, without the security risks that are common to the TCP/IP, since they are isolated from the Ethernet and IP domains. At the same time, this permits very easily the use of secure peer-to-peer communications between users of the network in the Ethernet domain.

The dual protocol stack allows communications with remote systems using the connectors for the Ethernet domain, or for IP over Ethernet. The network borders are the access points of the service,

and they will be offered in the Ethernet domain with the IEEE 802 protocols and the IP domain with IP, UDP, and TCP.

Each TUE will be able to work on the secure zone of the Ethernet domain and use, without risk, the resources offered by the UETS network servers: applications, data, audio, video, control, etc., and at the same time control network congestion, collaborating with the UETS central offices and accessing the Internet using the non-secure IP domain.

Using the Ethernet stack on servers has significant advantages [1], including avoiding the congestion produced by the TCP and IP protocols, which are software elements; thus, it is possible to offload the communications processes to communication devices of the front-end type, since they are link-layer protocols.

Running Internet Applications: The Supervisor Module
The Internet model resolves the concepts of network services and terminals using a universal presentation, based fundamentally on the navigator model (e.g., Netscape, Opera, Internet Explorer) and on the specifications of W3C. The main problem with TCP and IP is that they correspond to layers 3 and 4 of the communication architecture reference model, which are software applications; this imposes a limit on efficiency on both the hosts and the routers. We can take as a reference the strategy used by virtual circuit networks—when the X.25 protocol at L3 made it technically impossible to increase the network connection speed, the switching had to be carried out at L2 with frame relay.

In the case of Internet, it would be a basic necessity to adapt the applications so that instead of the IP protocol, they would use MAC 802.3 frames directly; their six-octet addresses have 281,474,976,710,656 possible combinations. While IP datagrams have to be delivered using routing procedures, the MAC 802.3 frames use the new physical switching mechanism of the UETS central office, which is absolutely necessary in order to offer the advanced network services that are proposed herein.

Common Internet applications will be executed under a supervisor module that will establish communications on the Ethernet domain using the MAC/802 protocol stack connectors, and on the IP domain using the TCP/UDP/IP connectors. The Ethernet and IP domains can co-exist in the same physical infrastructure, since the terminals described herein are able to operate over both indistinctly and at the same time.

The boot-up process of the terminal is immediate, since the terminal only has to load a small program. If there is any problem, it would only be necessary to turn off the terminal and turn it on again, since the browser program and the file handler are located on a non-volatile memory. If an irrecoverable failure occurs on these programs, the terminal will include mechanisms to carry out a complete bootstrap from the application repository servers located at the service provider facilities. The service provider will also be in charge of updating the terminal software when needed. A solution could be that every time the terminal starts, it would establish a connection to the application repository server, which in turn would send the application updates or the complete system if needed.

The Ethernet Central Office (CUE)

This device will connect the users of the service by using any kind of access media. The Ethernet domain is constructed by interconnecting central offices, as shown in *Figure 5*.

The characteristics making this system different are frame switching—which occurs at the physical layer, allowing for very high speeds—and the capacity to carry out error correction, plus flow and congestion control with the collaboration of the terminals. Using MAC addresses to route the traffic, the service offered corresponds to a L2 virtual private network; the number of terminals could be several orders of magnitude bigger than the current Internet [6]. Since MAC addresses could not be faked, any potential attacker would be identified almost immediately.

The physical switching and the mechanism described allow for a high degree of simplification of the equipment, which uses only hardware switching using Banyan switches. On the other hand, the connections on the Ethernet domain are almost immediate and do not require any signaling, so there is less overhead and an increase in bandwidth for not using TCP/IP, and it is easier to offload the communication processes to specialized front ends.

Another difference with IP routers is that the traffic and congestion control, error detection, and

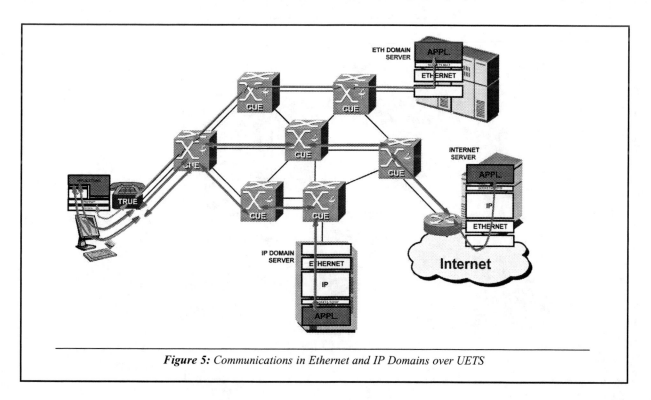

Figure 5: Communications in Ethernet and IP Domains over UETS

correction are done in co-operation with the TUE terminals and the CUE central offices using link-layer procedures throughout the physical network; this guarantees the bandwidth and the latency needed by the applications, only bounded by the existing capacity, as there is no mechanism able to "get something from nothing."

Multiple-Play Services

The telecom industry is trying to develop solutions to feed the insatiable appetites of the consumer and enterprise markets for voice, video, and data communications on a single network. To the global broadcast and telecom industries, the triple-play opportunity is at once a potentially inexhaustible source of revenue and a constant source of frustration. Despite the huge earnings for successfully delivering multiple-play services, there remain considerable obstacles that continue to make access difficult. The following section describes several application scenarios of the UETS technology in multiple play, although it is good to keep in mind that there are almost unlimited possibilities. The types of services that will be offered in UETS are described in *Figures 6a*, *b*, and *c*.

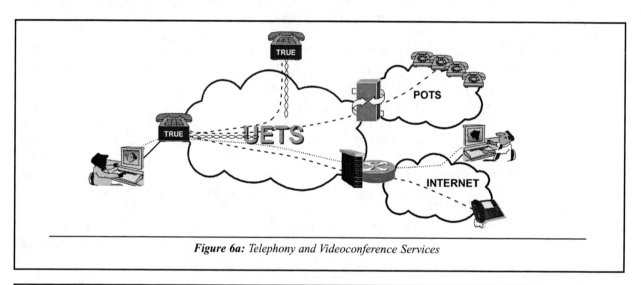

Figure 6a: *Telephony and Videoconference Services*

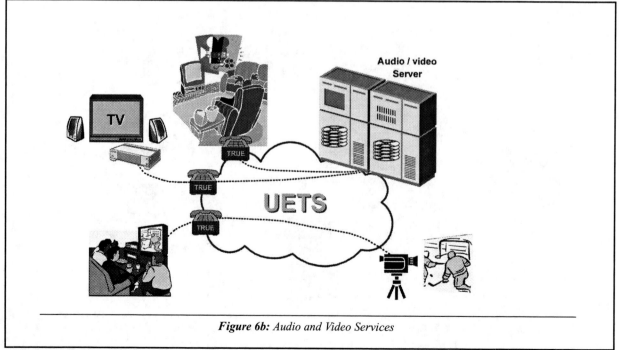

Figure 6b: *Audio and Video Services*

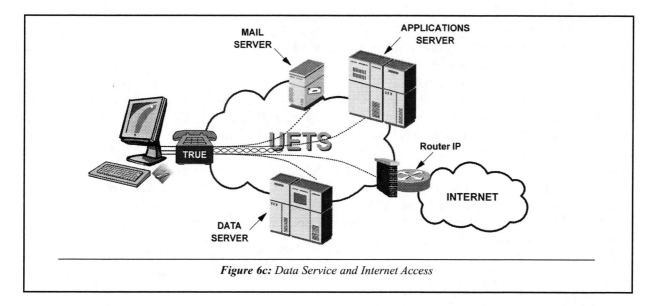

Figure 6c: *Data Service and Internet Access*

Broadband Services

One of the main problems preventing the widespread use of broadband services is the need to connect through a PC, which has an operating system; this is excessively complex for the majority of the potential users of these services, as they are not able to understand how it works or maintain it correctly. Additionally, we should consider the high costs of investing in these systems, as well as their short operative lifetime.

There is no need to describe the severe security risks caused by viruses and other types of attacks coming from the Internet. The average user feels helpless regarding both of these security risks, which will cause problems sooner or later, and the difficulties involved in managing data backups and the information stored in the computer. The TRUE terminal solves these problems completely, as professional people are managing the system.

In the first place, as the TRUE terminal does not have an operating system, most of the attacks coming from the network are minimized, as the terminal is connected to the Ethernet domain, a protected environment controlled by the service provider. As the local network is extended to the service provider, the provider will be in charge of installing, managing, and controlling the servers: application, data, e-mail, name and address, audio and video, etc. With this system the software-licensing problem is solved: The user pays for the service, and, in turn, the service provider pays for the licenses.

When dealing with audio and video, something similar happens, as the users will pay for the access to these contents. With these new services, the telecom network operators can compensate for the loss of revenue from telephone billing, caused mainly by the expansion of VoP.

The end user could have servers installed at his or her location, and all the traffic would be restricted to that network segment, since he or she is connected to the network services provider using TRUE, which is able to filter the traffic directed to the access network.

Multi-Processor and Storage Systems

The system can also be used to build supercomputers or advanced network equipment, using multiple parallel processing and storage systems connected at Gigabit speed using the UETS central offices. They offer very high-speed physical switching between processors, using the direct connectors to the MAC/802, making it possible to combine both techniques to guarantee the required bandwidth for critical communications, congestion control, and error detection and correction.

For inter-process communications (IPC), the protocols stack with Ethernet connectors is employed, while the connection can be gigabit or 10 gigabit Ethernet. Using this system, it is possible to create systems with several thousand processors in parallel.

Business Applications

The universal Ethernet service is effective for enterprise networks, which can use it in many ways, applying the "computer on Net" model [1]. It minimizes the per-seat cost, increases the lifetime of the terminals—as they are only a screen and a keyboard—centralizes the management, and lowers the security risks, since there is no operating system to attack and the supervisor programs can only be modified by the service provider. Big companies or organizations could install and manage their own internal networks with their own resources, outsourcing those parts that would be most cost-effective on the service provider network.

The UETS system makes it easy to create a paperless office, as the users do not have to work on a specific terminal or workstation. All the information they need is stored in the servers, which are accessible from any point of the Ethernet domain, which, by means of its characteristics, offers a global coverage.

The possibility to connect directly to the telecom operators and through the power lines with PLC technology solves the backup problem to keep the service operating in case of a power failure. This also allows a high degree of connectivity and availability of the service, as a complete failure may only occur if both the power and telecom access fail.

Replacement of PBXs

The private branch exchanges (PBXs) used by businesses could be replaced directly by a CUE with voice-grade twisted-pair output interfaces for the extensions already installed and replace the telephone terminals that need data connection with TRUE terminals, connecting the data equipment on the corresponding output. Migration could be carried out from a voice-only installation to a data and voice network over the same cabling.

Mobile and Wireless Services

The integration of wireless and mobile services based on packets over the proposed network is almost immediate. The Ethernet domain is the ideal medium to interconnect the general packet radio service (GPRS) or Universal Mobile Telecommunications System (UMTS) access

points, and also the wireless access points based on IEEE 802.11/Wi-Fi, IEEE 802.16/WiMAX or IEEE 802.20.

In those cases where the original framing is done with IEEE 802.3 frames, as in Wi-Fi, the service integration is transparent. In the rest of the cases, it could also be possible to use this framing, which might be advantageous for mobile systems on packet mode, since this minimizes the overhead of the protocols and optimizes the communication efficiency through the wireless medium.

UETS supports real mobility for users, devices and services, and access to operator services anytime, anywhere. Users can take their home Wi-Fi telephone and use the phone at Wi-Fi hot spots to continue to receive calls. With UMA, this service can be extended into the GSM network.

Mobile users can be identified by user login and password, registration address, IP address, mobile station integrated services digital network (MSISDN) (normal telephone number), International Mobile Subscriber Identification (IMSI) number (numbers from mobile cards), UETS address, or UMAC address.

Energy Saving

UETS technology would be included in a future power saving strategy, due to the potential environmental benefits of its interaction with the mains power supply grid to build the "intelligent network." One of the most important aspects of generation systems based on renewable energies is the temporal correlation between demand and generation, because they change the basic concepts of conventional generation systems. The key to taking advantage of these resources is the adaptation of demand to supply (demand management) and not the opposite. Here resides the high potential of an integrated or convergent approach to the electricity and telecom networks.

The system described could be used in a highly advantageous manner for systems that optimize energy savings. Through the use of the 802.3 frame format on the user's network, adding terminal functions to household appliances, the electric company could control these devices, activating or deactivating them as needed by the power grid load and electricity distribution systems. In this way, during the summer, as the temperature rises, the air-conditioning system thermostat could be raised to lower the power consumption, washing machines and dishwashers could be activated at low-usage times, and so on. This would lower the user's bill and improve the usage of power plants and the electric company's power grids.

Conclusion

UETS is a new switching architecture that offers complete scalability, enhanced L2 security and high performance with simplicity in the switches. It is based on the hardware switching of Ethernet frames using topological and hierarchically assigned standard local MAC addresses. The architecture has very good performance-to-cost ratio and is compatible with existing Ethernet and IP networks.

The UETS system makes it unnecessary to migrate the current Internet over TCP/IP, since the two technologies can co-exist indefinitely and harmoniously. The current Internet is another service of the UETS that employs the TCP/IP stack. The UETS users can access their data or services inside the Ethernet domain in a more efficient and secure way, but at the same time, there is full connectivity to the IP domain. From that point, the user can access the universe of Internet services. How both domains will evolve depends on what the providers offer and what information the users access, since from a technological point of view, no migration is needed. Both domains can co-exist and work in a complementary way, as opposed to what happens with IP version 4 and 6, which compete with each other.

Ethernet offers link-layer and network IP, while IPv4 and IPv6 offer network services. UETS technology corresponds to the link layer, which supports any kind of network; it is multiprotocol, per se, and an ideal way to connect terminals throughout a single network and not via a set of interconnected networks.

The main advantages of UETS architecture are its inherent security, wire speed performance, compatibility and interoperability with existing IP networks and applications, and lower cost-to-performance ratios. Applications specially suited to this architecture are high-performance provider networks of any size (e.g., LANs), multiple-play access networks, scientific networks, distance learning, computing on Net, HDTV distribution, home networking, storage networks (storage area networks/network access servers), military secure networks, networks of workstations (NOW), and L2 VPNs.

References

[1] Jose Morales Barroso, "From Computer Networks to the Computer on Net," IEEE Communications Magazine/GCN, October 2005, pp. 2–4.

[2] IEEE 802.3ah. "Media Access Control Parameters, Physical Layers, and Management Parameters for subscriber access networks."

[3] IEEE 802.3ah EFM, "Comments received during LMSC sponsor ballot" [on-line]: www.ieee802.org/3/efm/public/comments/d3_1/D3_1_proposed_responses.pdf.

[4] Jose Morales Barroso, Guillermo Ibanez, "Ethernet Fabric Routing," IEEE INFOCOM. High-Speed Networking Workshop: The Terabits Challenge. April 2006.

[5] IEEE STD 802-2001, "IEEE Standard for Local and Metropolitan Area Networks: Overview and Architecture."

[6] Jose Morales Barroso, "UETS: Towards a new layer 2 based Internet." Exploiting the Knowledge Economy: Issues, Applications, and Case Studies. P. Cunningham and M. Cunningham (Eds.) IOS Press. 2006, pp. 1615–1622.

Bundles and Range Strategies: The Case of Telecom Operators

Sophie Pernet

Project Manager, Consumer Market Services
IDATE

Abstract

Against a background of competition and the generalization of IP that characterizes the field of electronic communications, the concept of the "bundle" has resulted in the emergence of triple play and quadruple play. This paper offers an overview of the growth of this phenomenon by introducing a distinction between the basic components of multiplay strategies and the diverse range of functions that can be linked to these strategies.

Introduction

Bundle Development in a Context of Fierce Competition and "IP-ization"

Telecommunications operators are experiencing a dramatic change in their market, with competition growing increasingly fierce on the one hand and a technological shift marked by the generalization of Internet protocol (IP) and the growth of very-high-speed networks on the other. The decline in the value of the traditional markets for fixed telephony is continuing as a result of tougher competition but is primarily due to substitution by voice over IP (VoIP) and mobile services. The broadband market, which has facilitated the emergence of new players (Internet access providers), is booming but is characterized by competition that can sometimes be very fierce, which has a negative influence on profit margins and has obliged players to increasingly invest in infrastructures. Last, the pay TV market is approaching saturation in some countries, audiences are increasingly fragmented, and the launch of new forms of access such as asymmetrical digital subscriber line (ADSL) and digital terrestrial television could weaken the position of the sector's traditional players.

In this context, the objectives of bundle strategies in the electronic sector are classic, namely to increase average revenue per user (ARPU), ensure consumer loyalty, and expand subscriber bases by introducing a differentiating factor.

The Influence of National Contexts

The competitive context of each market is a key factor in the commercial strategy pursued by operators. Regulators are sensitive to the risks that can be linked to the launch of a bundle by a player that uses its dominant position in a service to finance its move into a new market. That has nevertheless not prevented the growth of bundles, a phenomenon that can even be seen in highly competitive markets where the most bundles are to be found. In the United States, regional Bell operating companies (RBOCs) rushed to offer consumers long-distance local flat-rate bundles after the Telecom Act and are now using bundles as a weapon against cable operators by marketing triple- and quadruple-play offerings.

In France and Japan, very fierce competition in the broadband market has turned these countries into highly advanced markets in the coupling of ADSL access services, IP telephony, and television. On the other hand, in markets that are not very competitive but still enjoy a steady growth rate, incumbents are not showing the same eagerness to enhance their offering by launching VoIP, for example.

However, given that competition in this sector is bound to become widespread, it is very likely that multi-play offerings will become more common.

Two Bundle Strategies

Bundles cover a range of relatively different realities, which we propose to distinguish between by identifying the following:

- The horizontal bundle, double, triple, or quadruple play is aimed at linking relevant services from different markets such as Internet access, TV, and fixed telephony or mobile services. These bundles are part of a broad-range strategy.
- The vertical bundle consists of linking additional functions to a reference service such as Internet access, TV, or telephony. These bundles are part of a deep-range strategy.

Broad-Range Bundles

Triple Play Heading toward Generalization; Quadruple Play in Sight

Although this does not eliminate the risk of foreclosure, bundles in the communication sector are remarkable in that they force players in relevant markets, which are in principle different and separate, to compete against each other. As we have seen, on the side of traditional telephone operators are Internet access providers, cable operators, and now mobile operators. This group also includes virtual operators and distributors that initially specialized in one market and are now oriented toward bundle offerings, either to compensate for revenue losses in their initial core business (the resale of minutes of telephony) or in response to competitors offering multi-play.

Each player has thus sought to position itself in new markets: Internet access, VoIP, then IPTV for incumbent operators; voice and IPTV for access providers; and voice and Internet access for cable operators. As a result, fixed and mobile telecom, access providers, and cable operators are likely to find themselves competing head-on in the three dimensions of quadruple play: fixed and mobile telephony, Internet access, and TV in the consumer market.

An operator's strategy depends on several factors, but especially on its positioning in the market and its marketing objectives. Regardless of the type of player in question, however, the orientation toward quadruple play is a major market trend. Almost all incumbent players offer quadruple play, and challengers such as Internet service providers (ISPs) or cable operators plan do to so.

Video on Demand or Television: Different Visions of Video Integration in Bundles

The basic components of triple-play offerings of any group of operators are likely to be fairly different. This is true of the telephony service, which can be offered on an analog or an IP basis, and at special prices for international calls, as well as those to mobiles. This is also the case for Internet connections depending on connection speeds and prices. However, the options tend to vary the most for the video component. Depending on market conditions, triple-play operators will either offer existing bouquets traditionally distributed by satellite or make up their own bouquet by integrating "traditional" television channels from free digital terrestrial TV or pay TV. In some cases, when access to television channels is difficult and controlled by a powerful player that possesses its own distribution system (i.e., the United Kingdom), operators may focus on launching a video on demand (VoD) service.

Are We Heading toward Quadruple Play?

Incumbent operators are obviously in a better position to offer quadruple play since they usually have a mobile operator as a subsidiary. However, cable operators are also in the running, including Virgin Media in the United Kingdom, which already offers quadruple play.

However, the quadruple-play offering is still in its infancy. The first offerings are still often made under different brand names and offer a low level of functional integration. As a result, operators rarely offer unified services related to quadruple play such as unified messaging. This depth of range should increase with the diffusion of offerings and greater market maturity.

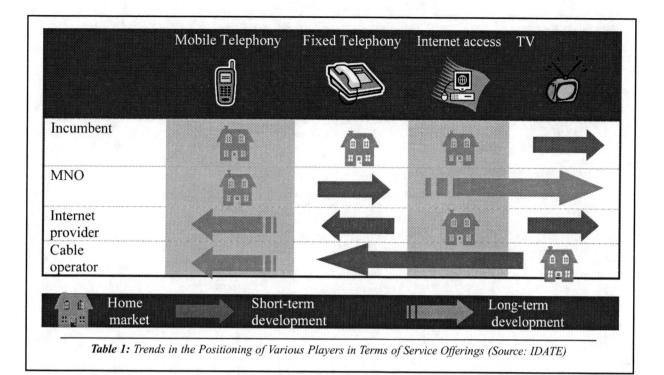

Table 1: *Trends in the Positioning of Various Players in Terms of Service Offerings (Source: IDATE)*

Several Technological Orientations for Quadruple Play

Quadruple play, based on the addition of mobile services, can cover a large number of scenarios.

The most classic model is that of adding cellular services to the triple-play offering; this is the most widespread form of quadruple play, which may or may not bear the brand name of the cellular operator. Operators can also sign mobile virtual network operator (MVNO) agreements with the latter in order to distribute services under its own brand name (this is one possibility envisaged by Free in France).

However, mobile services can also be offered in the framework of fixed-mobile convergence (FMC) services, integrating an element of interoperability between several basic multi-play components. BT was a pioneer in this field with its development of an offering called BT Fusion. This offering is aimed at providing wireless telephone communications, which either use ADSL networks via a wireless connection or cellular networks if the user is not near a point of access. The innovative nature of this offering lies in the fact that it relies on a single terminal for both mobile and fixed communications, which switches networks automatically according to the geographical location of the subscriber. Access points for IP communications are either a base sta-

tion in the subscriber's home or public hotspots. The advantages of this service for users include a single terminal that reduces costs (notable of mobile calls) and a single bill for the two services.

Mobile "pure players" seem to be threatened by multi-play developments. The latter could give rise to strategic alliances and a new cycle of mergers and acquisitions, even if other options are envisaged. As a result, some mobile operators are offering specific prices for mobile calls made from home ("homezoning"). This mainly pricing offering is not a bundle as such, but it acts as a substitute, competing with double-play offerings and discouraging the transfer of traffic to VoIP. Similarly, backed by the bandwidth speeds that their third generation (3G) and high-speed downlink packet access (HSDPA) networks can offer, they may be tempted to compete with ADSL to offer occasional and mobile Internet users performing connections.

Moreover, some mobile operators signed agreements with fixed operators to offer Internet access and fixed voice, including 02 in Germany.

Deep-Range Bundles

Deep-Range Strategies

In the search for avenues of growth, operators have

Country	Operator	Fixed voice	Internet access	Television	Mobile services
United States	AT&T	■	■	■ Satellite + IPTV	■
	BellSouth	■	■	■ Satellite	■
	Verizon	■	■	■ Satellite + IPTV	■
	Qwest	■	■	■	■
France	France Telecom	■	■	■ IPTV	■
Netherlands	KPN	■	■	■ IPTV	■
Sweden	TeliaSonera	■	■	■ IPTV	■
United Kingdom	BT	■	■	■ IPTV	■
Italy	Telecom Italia	■	■	■ IPTV	■
Spain	Telefonica	■	■	■ IPTV	■
Germany	Deutsche Telekom	■	■	■ IPTV	■
Japan	NTT	■	■	☐	■
South Korea	Korean Telecom (KT)	■	■	☐	■

■ Offered ☐ Planned

Table 2: Positioning of Incumbent Operators in Quadruple Play

long since associated deep-range services with a reference service such as Internet access, voice, or TV. Each type of operator has consequently worked on its range of services based on its core competence: convenience services (i.e., caller ID display, call waiting) for telephony operators, and channel bouquets, premium channels, pay per view (PPV), VoD, and digital recording for cable and direct broadcast satellite (DBS) operators. As for Internet access providers, they can find ways of differentiating their offering and boosting their profitability thanks to access-related services (e.g., antivirus, messaging services, security services, storage services, portals, related services).

Fairly logically, we can see that operators have a past to contend with. They generally have a range that is deep in terms of their core competence and fairly restricted in the new areas of triple play.

Cable operators, on the other hand, have long since offered a wide range of television services, but their offering of pay services related to access or VoIP is relatively limited. Incumbent operators, by contrast, have developed a large range of pay services related to access, but few services in the TV and VoIP markets.

Moreover, operators are tending to integrate a growing number of services in their base offering. This is notably the case with services related to VoIP, which are most often included in the subscription. This trend is also true of services related to Internet access. The integration of "free" services in the base offering consequently represents a way of offering a differentiated range.

What Are the Trends toward Deepening Ranges?
While operators are generally tending to enlarge

their range of services, the situation varies to a greater degree when it comes to the deepening of ranges around a basic service such as voice, Internet access, or TV. Several factors influence range deepening, including the following:

- Primarily, the quest for complementary revenues based on a highly competitive basic service; as a result an operator can advertise an attractive flagship offering, in access for example, and capture additional revenues. As a reference, convenience services for dial-up voice can still generate significant revenues for incumbents.
- Related services that also enable players to differentiate their offering from that of their competitors—especially in markets such as Internet access, where the basic offering is essentially characterized by bandwidth capacity and is therefore not differentiated. This is the case with BT, for example, which links the services of the Yahoo portal to its offering and thus positions itself differently from other ISPs.
- Along the same lines, related services enable players to build specific offerings for certain client segments. For example, in association with Internet access, Telefonica's Disney pack combines cartoons, parental control software, and educational games and is aimed at children. In mass markets such as voice, TV, and now Internet access, the growth of a segmented offering will be crucial to winning market share and targeting different types of customers effectively. The launch of targeted bundles of services is a key component of this strategy.
- The addition of premium services also contributes to the image of the operator's brand. As a result, some high value-added services contribute to the operator's image as an innovator. This is the case with video telephony, for example, which has notably been developed by France Telecom and Telecom Italia.
- Lastly, new services are appearing and naturally enhancing the service range. For example, the fact that VoIP is linked to a personal computer (PC) significantly improves the outlook for offering communication management services, and it is certain that new services will emerge.

All of these factors are going to encourage operators to develop a deeper range, especially in a fiercely competitive context. However, other factors should slow range deepening including the following:

- Primarily, the trend toward integrating services in the basic offering to maintain ARPU. In the field of access, ISPs unanimously agree that the basic offering tends to integrate additional services such as antivirus. Moreover their offerings already include services related to messaging (e.g., e-mail, addresses). In the competitive TV market, bouquets tend to be enhanced in order to avoid decreases in prices. Lastly, alternative VoIP operators generally offer free call management services, and it will be relatively difficult to make clients pay for these services.
- The launch of new related services also involves management costs and risks in terms of quality of service (QoS). The costs generated by these new services need to be offset by these revenues, which is not necessarily always the case. As a result, some operators may prefer to stick with a standard offering.
- Expanding the range also means looking for partners, notably in the world of TV. This can prove difficult, especially for small ISPs that have little bargaining power compared to the majors.
- Last, some operators may choose to market a very simple offering that is very transparent and enables consumers to find a suitable offering quickly and easily. The complexity of the options can constitute a barrier to purchasing, especially for new subscribers. Some players may choose to remain positioned in very basic offerings to make their offering very transparent, notably compared to incumbents, which generally market a relatively complex offering.

These factors tend to reduce the range of services offered, although it is possible to identify the following trends by type of player:

- Incumbents have R&D resources that enable them to develop innovative services, such as video telephony and converging services with mobile terminals. They position themselves more in terms of deepening their range based on access and should continue to deepen their range to differentiate themselves from new entrants.

- Cable operators are traditionally positioned in the TV market with offerings in basic services, premium services, thematic bouquets, individual channels, and VoD. They have proved slow to deepen their range in Internet access and telephony but will be obliged to develop services in the future by the competition.
- Lastly, ISPs have developed specific strategies. An operator like Free, for example, offers few services related to access, several paying services related to voice, and a large range of bouquets and optional channels for television.

From a market dynamics point of view, the features distinguishing telephone operators from Internet access providers should nevertheless tend to fade away. Similarly, the frontiers between cable operators, which have seen major consolidation in their sector and have often made major investments to update their networks, and telecom operators with optical infrastructures should become less and less visible. Players will nevertheless continue to differ in terms of their financial power, their competitive advantages, and the market sectors that they primarily target.

Conclusion

How Profitable Are Bundle Offerings?

Policies for price reductions on bundles vary significantly: some players, notably U.S. RBOCs, offer a marginal reduction, with the main advantage for users being a single bill. Other operators go as far as to offer a 30 percent reduction on a triple-play offering, with the "average" price reduction being around 10 percent. On the other hand, price reductions are also aimed at orienting consumers toward a product in the range, and two reduction strategies are offered: on one hand, some operators offer bigger reductions on top-of-the-range offerings to encourage customers to subscribe to their greater added-value offerings; other operators, on the contrary, offer attractive reductions on entry-range offerings in order to maximize their appeal, with the long-term objective of converting these subscribers to other services. Last, let us mention the case of Carphone Warehouse, which offers free Internet access to its dial-up voice subscribers.

Although the launch of bundle offerings may boost ARPU, it can also squeeze profit margins, especially if operators have to buy high added-value contents. However, the amortizement of the acquisition cost of subscribers over several services, significant savings in terms of technology and organization, and client retention should boost operators' margins. It is consequently extremely difficult to assess the overall profitability of bundles. Some experts remain skeptical of the potential of bundle offerings to generate a real increase in profits, and it will undoubtedly be necessary to wait for feedback based on the experiences of various players before making any concrete judgments. There would seem to be a few encouraging examples in the United States, where the bundling strategy enabled the major local telephony operators (SBC and Verizon), that had been suffering from a significant decrease in lines and traffic, to boost their turnover in the fixed market in 2004.

Variety and Transparency in Multi-Play Offerings

The bundle is a key aspect of the range strategy and will undoubtedly condition operators' success. Growth in the broadband market, the diversification and intensification of usages, and the growth of new services (i.e., FMC services, unified messaging, videoconferencing, HDTV, interactive TV) represent a wealth of opportunities for bundles.

Sticking to marketing strategy components and thus not dealing with complex execution problems, several risks and critical points, including the following, can nevertheless be identified:

- In markets where all players are positioned in triple or even quadruple play, the offering runs the risk of becoming commonplace, and a loss of references for consumers is to be feared.
- Operators therefore face the challenge of creating an original range of services, with strategic choices in terms of the depth and width of the offering. To date, players have primarily developed their range according to their core competence; they now have to do the same for the new markets in which they are positioning themselves to offer consumers a range of harmonized services. Operators will notably have to trade off between the richness of their range

and transparency: too large and complex a range could repel potential subscribers, whereas a simple and transparent range is limited in scope.

- In markets that are approaching saturation in some cases, bundles of services also need to evolve to address specific client segments. The challenge will then consist of identifying market target and designing specific offerings for each segment.
- The reactivity of players may be limited by hesitation on the part of regulators, especially as the same authorities do not always regulate all of the components of multi-play.
- Competition in multi-play is not limited to that which characterizes the different players in access (i.e., telcos, cable operators, access providers). It can be extended to all players in a value chain that is becoming increasingly complex and includes the distributors, Internet giants, and software players.

Main objective of the bundle strategy	Positioning of operator	Nature of dominant bundle offering	Operators pursuing this strategy
Win loyalty	Incumbent facing challenge	Deep and wide range	France Telecom
Boost broadband demand	Incumbent in dominant position	Relatively narrow range	Telefonica, Deutsche Telekom
Win market share	Challenger, new entrant	Low-cost bundle	Free, Fastweb
Increase ARPU	New entrant	Bundle oriented toward value-added services	Tiscali
Target new market segments	Cable operator	Wide range	UPC, Time Warner Cable, Comcast, Telewest
Win loyalty and attract new clients	RBOCs	Quadruple play	SBC, Verizon

Table 3: Typology of Bundle Strategies

The Thing about Customers

Thomas Jürgen Quiehl

Executive Vice President, Innovation and Technology Strategy
T-Com

Overview

This essay fervently makes a twofold appeal. On one hand, it aims to encourage those involved about the inherent potential in the services yet to be developed in the convergence area. On the other hand, it aims to emphasize the merits of re-engineering productization by shifting the focus from technology to genuine customer segmentation.

If we examine the current topic of quadruple play and the gradual derivation of further concept strategies from it, we get a consistent picture of a high-value agenda for innovation.

What Is Quadruple Play?

We could almost believe the market is trying out some fun game from our childhood days—single play, dual play, triple play, quadruple play, multiplay. That sounds a lot like a game of hopscotch with a lot of fields to be mastered. So what is it that this "game" wishes to teach us and, more important, what does it intend to teach the customers who will hopefully pay to play it?

You will remember that "convergence" was the innovative goal of the communications industry back in the late 1990s and that the term TIMES (telecommunications, information processing, media, entertainment, security) was for a long time synonymous with the idea of the world growing closer together. The term is as unwieldy as the technological challenges it stands for are formidable. The ideal scenarios did not manifest themselves, and TIMES vanished.

In its stead, two strains of innovation later emerged, moving the theme of convergence ahead in minuscule steps. Fixed-line telephony and the Internet/Internet telephony started to merge. Skype presented itself as a dynamic wave, and there, then, was the first version of the game. In recent times it has expanded to triple play, given expansion of the communications spectrum to include the television/entertainment field. The other development was the start of wrangling between fixed-line and cellular telephony, a phenomenon referred to either as fixed-mobile convergence (FMC) or fixed-mobile substitution (FMS), depending on the tone of the rivalry. And now both these strains of innovation are coming together to create quadruple play.

As a marketing term, quadruple play covers the services of television, fixed-line telephony, cellular telephony, and Internet. The network integration of voice, data, and video originally addressed in triple play is not experiencing expansion in terms of services, but is only seeing mobile distribution.

In a nutshell, quadruple play stands for the promise of the following:

- Universal integration of connectivity
- Mobilization and portability of use
- Enhancement of applications through enabling services
- Optimization of utilization costs

In competitive terms, the 1-2-3-4 game is due to the commoditization of basic communication or market saturation. Fixed-line providers, Internet service providers (ISPs), mobile operators, and even

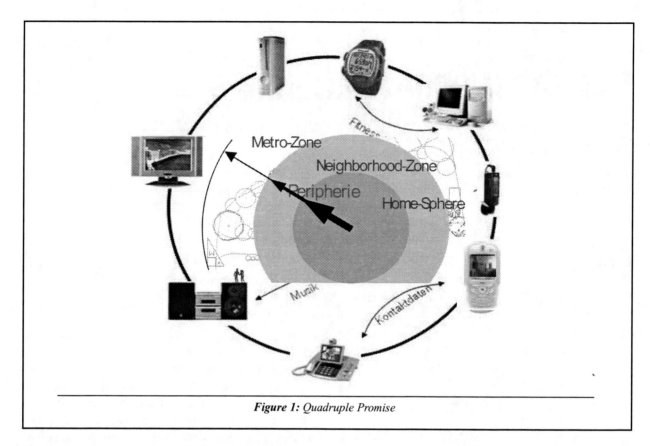

Figure 1: Quadruple Promise

media/entertainment businesses are impacting on adjacent value-creation fields. Additions to the range of services are aimed at (again) increasing customer benefits and compensating for the decline in margins.

What Do Customers Want?

Let us look back in time to the dim and distant past. Our hunting forefathers would have welcomed the ability to share information with their co-hunters or to rave about their day's successes back at the home cave. They would have appreciated a universal localization overview over the grounds and the ability to make use of the experience gained in previous hunts. They would also have loved to spend a quiet moment in the evenings to go over the events of the day in their minds.

The human race, and with it the technology it uses, has advanced. The activities that used to fill our lives have transformed to become extensive areas of utilization. Telephony (fixed-line or cellular), Internet (mostly personal computer [PC]–based), and television (primarily with an entertainment focus) all reflect the areas of utilization for the con-

sumer and the sectors and branches of industry that have evolved as a result.

The fundamental needs of comprehensive and situational communication have remained the same—the chief aspects thereof being as follows:

- The bridging of space
- The disengagement of time
- The supplementation of information and knowledge
- The inter-linkage of functionalities and applications
- Fun and entertainment
- Simplicity and security of use
- The perceived price

If we wish to seriously determine the relevance of innovative offerings for the customer, we need to take a step back and ask ourselves the question in the comprehensive context of trends and customer-segment profiles.

The picture presented by a trend is as diverse as the message it contains is clear: Life complexity for the individual is increasing dramatically, and the search

Predominant	Media in Everyday Life
Society	Dynaxity World
	Aging Society
	Female Shift
	Persistent Insecurities
	Digital and Social Divide
Individual	Empowering the Individual
	The Mature Consumer
	Simultaneous Living
	New Community Building
	Seamless Entertainment
	Staying Healthy

Source: DeutscheTelekom

Figure 2: Consumer Trends

for ways to manage and balance it out is growing at the same radical pace.

The presence of media and universal entertainment offerings, for instance, is an option for life-long learning, relaxation and recuperation, and individualization, as well as an opportunity for relevant services. These opportunities must be exploited. By the same token, however, we can identify an information overload, breaches of privacy, and seduction against which we need to guard ourselves.

The global development of society and the individual will lead to greater opportunities and options for us. This will require greater personal responsibility and necessity for decision-making in ever more complex spheres of our lives. While this may be inspiration for some, it is daunting and even a cause of anxiety for others. In addition to the challenges posed by global events, changes in society, and the huge selection of information and media offerings to choose from, an increasing degree of technology is infiltrating our lives. Until now, the claim that "more technology equals an easier life" has only proven true to a limited extent. In most cases, the price we pay for more simplicity in one area of our life is a disproportionately high degree of complexity in another.

A very clear transformation in customer behavior is ongoing on the Internet. In the marketing world, this phenomenon is known as Web 2.0. The Web 2.0 movement is centered on stronger user contribution and a more participative form of social interaction. Phrases such as "democratization of the Internet," "people's economy," and "swarm intelligence" highlight the direction this development is taking. Admittedly, the movement is only being embraced by a selected few right now (mostly those with an affinity for the Internet). But further development will spread out across the population as a whole in the coming years, inducing an economic and social impact that is not yet foreseeable. New-fangled terminology such as "prosumer" (the self-producing consumer) or "from broadcast to customcast" (the new form of user-generated content) indicates a clear change in customer demands—demands that will be increasingly hard to satisfy based on "business as usual."

The information and communication industry needs to be aware of and accept challenges of this kind in the real-life customer environment. Only if service offerings are genuine troubleshooters, if they fit into the respective context, and if the bottom line for the consumer is genuine (net) convenience, will

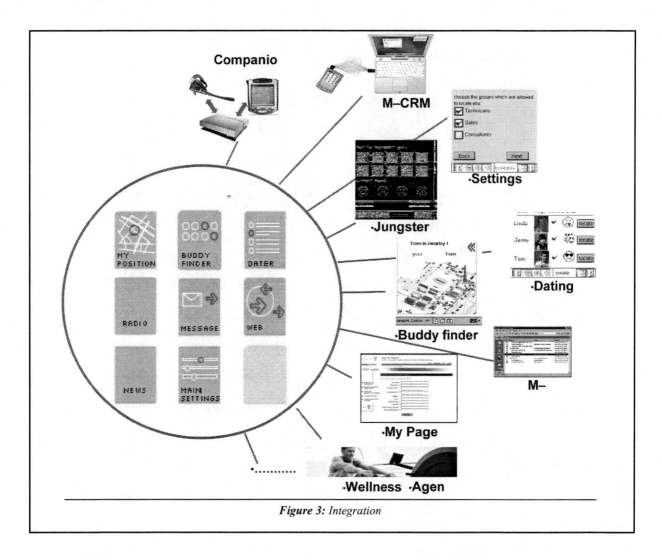

Figure 3: *Integration*

they be rewarded over the long term by consumer willingness to pay for them.

This is difficult for providers, since the traditional approach of primarily technology-driven portfolio development (i.e., the belief that more technology equals more benefit equals more willingness to pay) is coming up against boundaries. Cell phones, software programs, and consumer electronics, the offered functions of which only a fraction of the people they are made for actually understand and use, are some popular examples. You may frown at such an obvious example, but we have yet to see a movement on the market to counteract this continuing trend.

Another reason the situation is difficult for providers is that droves of customers formerly made up a homogenously uniform profile, but that standard profile is now disintegrating into a large number of very diverse segment profiles. Each of these segments can only be effectively—and, more important, consistently—addressed using profile-specific offerings (product and marketing). There are also the country-specific profile designs that ensue as services become globally available. Again, there is nothing new about that. And yet the early warning signs—such as the emancipation of users (Skype as an alternative option), user-generated content, and refusal by entire groups of the population to accept specific offers—seem, on the whole, to be made only marginal use of in portfolio development.

The Current Offering

Let us use the customer aspects we have portrayed as a filter on the quad-play offerings we can identify on the market—one step along this path of numbers at a time.

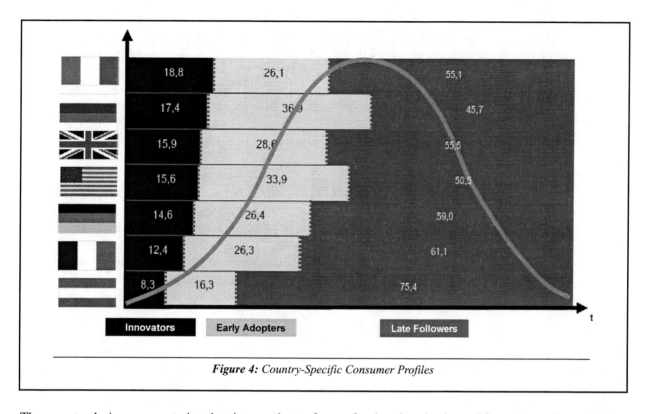

Innovators | Early Adopters | Late Followers

Figure 4: Country-Specific Consumer Profiles

The most obvious aspect in the integration of telephony and Internet has been the technical function of the access and the increasing pace of bundled rates becoming cheaper. This course is being taken one step further to triple play through the inclusion of TV access—primarily a technical, but also a price-relevant aspect to date.

Well, you think, integration is really only taking on any momentum with the advent of voice over IP (VoIP). But is it indeed taking on momentum? Okay, so the penetration of Skype services and the speed with which user devices have been generated have caused us to slowly detach ourselves from the PC as the necessary terminal device. Sadly, issues such as cross-provider interoperability or security functions have remained just as unresolved as those issues concerning the portability of communication across various devices or the personalization of use that IP technology would make possible.

Triple play—having only just advanced from infancy to toddler sophistication—is bringing television to our homes via the Internet on cable or telephone lines. A fantastic range of services is conceivable as a derivative thereof—the universality of communication across phone, PC, and TV; the integration of

professional and private life settings; the optimization of private workflows; and much more.

And what are the indications on the market? There may be advances in the diversity of television programs in some countries. And a lot of activity with regard to the distribution of content rights is a good thing too. This does not, however, constitute a qualitative leap. Where can we actually see a convergence of the Internet reservoir of information and content and the programs communicated via the output device that is television? Where can we see a follow-me effect on applications and content without the need to procure, install, and operate additional hardware? Not to mention the interactivity and self-generation of content or seamless communication.

Not surprisingly, we can expect to see quadruple play or the continuation of FMC in the form of mobile Internet access and mobile television. Mobility in society and a nomadic professional life will make FMC inevitable and sensible. The only thing is, is this next big service option being adequately addressed with the simple transfer of existing offerings to a mobile channel? Certainly not.

Where are intelligent services that converge localization, situational context, communication, and social interaction to relieve the user of tiresome searching, to allow spontaneous exchanges with our buddies, or to enable remote access to what the camera is capturing in our child's nursery at home? Where are the offers that enable ubiquitous transactions and payments across providers and voice portals, Internet sites, mobile interaction, or interactive points of sale?

In summary, the current triple and quad plays offer customers the following:

- A technical integration step
- The transmission of familiar content via a new channel
- Additional price reductions

Despite all this whining, however, it is undeniable that there is substantial technological and process-related complexity inherent in the described development paths. Creating seamless services across platforms, companies, and countries is a formidable task. From an entrepreneurial perspective, it makes sense to take a prudent approach and make small steps. And it is understandable to limit activities to "familiar" aspects so as to minimize risk.

There is the considerable risk, however, that offers are developed that do not actually meet customer demands—and this is the reason for the rather critical portrayals in previous sections. We have all seen enough examples of such misguided developments in recent years: wireless application protocol (WAP), Universal Mobile Telecommunications System (UMTS), and video telephony, to name but a few. In most cases, these failures were caused by the dominance of technology and misconceptions about application needs.

The current combination of an extremely difficult competitive situation on the market and the high degree of inherent potential in modern technologies would seem once more to have created an environment that encourages such misguided developments. And this is topped off by a reigning short-windedness of approach to entrepreneurial activity (manager targets, pressure from the capital market, and disruptive innovation can all be causes), with more weight given to short-term, somewhat standardized strategies than to sustainable business models, which tend to involve long-term commitment.

The consequence for individuals, businesses, and society is insufficient exploitation of the enormous potential of convergent and "life-optimizing" as well as value-creating services, which has yet to be tapped and adequately sold.

Where Do We Need to Go?

Allow me once more to address the troubleshooter criterion as a point of reference for market-relevant

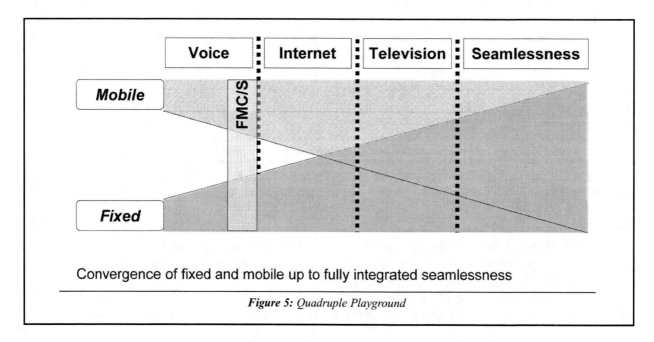

Convergence of fixed and mobile up to fully integrated seamlessness

Figure 5: Quadruple Playground

services. We can throw the different trends, signals, and market-research findings into the air like a juggler to create a structured interplay and then conclude the following core demands for communication services in the wider sense:

- The user increasingly does not want to (or in some cases cannot) deal with technical connectivity. Viewed independently, fixed-line phones, cell phones, cable, Bluetooth, wireless local-area networks (WLANs), and all the other options out there now and in future are about as interesting as a key to a door. The demand we can conclude is that the promise of seamless connectivity or ubiquitous access be kept and all complexity become self-organizing technology that silently operates in the background.
- In both private and professional environments, the different information systems (e.g., e-mail, news/RSS, search functions) and functional applications (e.g., shopping, taxes, housing, leisure) are important to the user. The next demand is intuitive use. This starts with portability on devices or in utilization contexts, includes security and privacy elements, and does not stop at personalization and individualization as well as context sensitivity. And all relevant application fields in an increasingly integrated life need to be covered and be made scalable in terms of their performance features. Lending a functionality term from the world of IT, we could describe such service types as ambient services.
- And the "fun factor" also has an important role to play. This does not refer so much to the type of content and entertainment offering as to making user interfaces and designs as emotional as they are functional. It is also important to enable social aspects such as building community, sharing information, meeting people, experiencing things with others, portraying oneself, and much, much more. The products offered by carmakers, for example, combine functional and fun elements. They show us that this makes sense.

The thing missing on this list is price. Economic considerations are also relevant for buyers, of course. The newfound appeal of thrift illustrated by such slogans as "Geiz ist geil" ("Stinginess is

cool") and the success of discount chains speak a clear language. The simple demand for cheaper products and services usually, however, applies to services, the "troubleshooter character" of which tend to be limited or which are "hygienic" factors in the eyes of the customer. When it comes to "sophisticated" (i.e., troubleshooter) offerings, price perception is different, as is people's willingness to pay the price. Customers are willing to pay considerable amounts for a wristwatch worn as a status symbol (in much the same way as we might pay a lot for a car, home accessories, or apparel), even though cheaper alternatives are available. Personal health and fitness is another area in which people are willing to invest a lot of money.

As a consequence, we need to take our deliberations on single, dual, triple, quadruple, and multi-play out of the technology niche and move them to the customer-segment-specific "troubleshooter role." Successful offerings (i.e., high-value business offerings) exploit potential by creating an integrated overall system of modular, inter-functional, scalable services.

This kind of ecosystem approach is based on technology and operational performance, principally defining itself, however, contingent on customer-segment-specific demands and translating technology into utility and benefit. Fortunately, we are able to break down such a demanding procedure into manageable portions and apply an analogy with nature, since the earth's ecosystem, too, comprises cascading ecological systems in specific biotopes. Providers, therefore, are not required to "make the whole world happy" all at once, but can let the message in their business context, for service clusters, terminal devices, interfaces, and content offerings, become reality for the customer in a gradual process. Allow us to use the much-cited example of Apple. A terminal device (the iPod, a concept focused fully on a specific use) was the basis from which an integrated utility developed (iTunes), with the next integration wave already foreseeable (iPhone).

But enough about the holistic classification and description of strategic impetus for the innovations agenda in the environment of communications services. What examples could we come up with for customer-centered eco-system offerings?

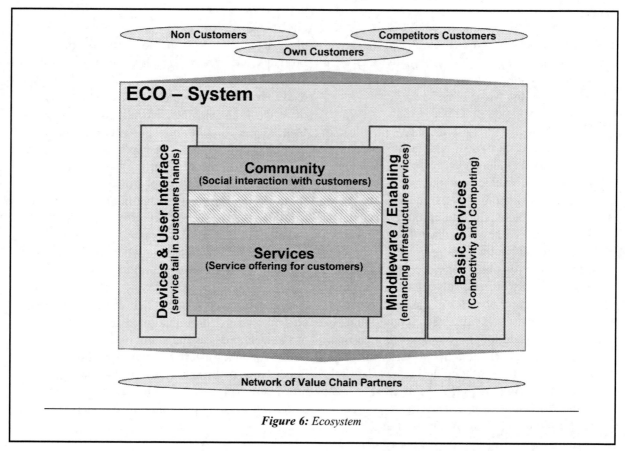

Figure 6: Ecosystem

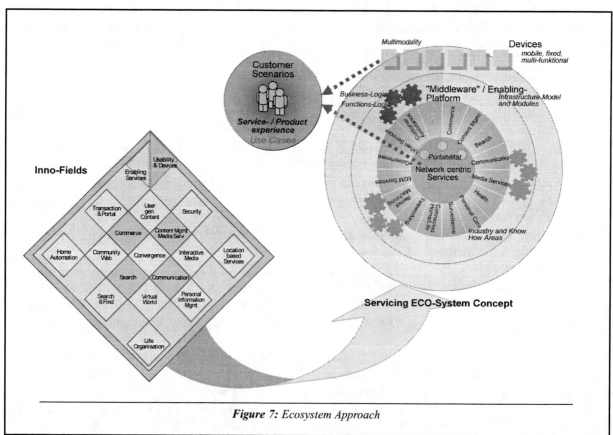

Figure 7: Ecosystem Approach

Voice Services/One Phone

Voice communication remains a central application. The one-phone concept (also referred to as a single phone in ongoing market projects) is to make voice communication technically limitless and justifiable in terms of price. Technical integration and portability across all devices and fields of use is a first step worth pursuing, though this does not mean all is mobile airtime. The long-promised universal integration of address book, buddy list, voice-mail box, and profile-based billing would be a usage-oriented move. And the implementation of voice-controlled services for input and output could have a usage-optimizing effect. Reading out or speaking text messages, controlling functions using voice commands, or the multi-modality of output (analogous to hands-free navigation with spoken commands) are simple elements of this. Voice communication could become genuinely seamless if a video call received by cell phone were to automatically send its voice component to the mobile handset and its video component to, say, the television or PC when we approach a display device.

Data Services/Universal Archive

End-to-end usage, ubiquitous access, and context-specific mapping of roles are elements in quad play. It is in this context that universal archives are predestined to be a network-centric platform service, to act as troubleshooters. The constant fear of losing the digital images we store on our PC and the tribulations involved in social integration á la Flickr and YouTube could be remedied using virtual memories.

They could also be used to perform constructive information life-cycle management, possibly even automated via semantic algorithms, with data being structured and secured according to relevance. Working in groups in both professional and private contexts (e.g., the joint compilation of a journal or scrapbook) is possible using shared workspace and identification-based access mechanisms, even for target groups with a lesser degree of affinity to on-line operations. Automatic format changes (version changes for digital media) and rule-based reminder functions (alerts for deadlines taken from filed insurance policies or warranty agreements) would also be possible using such archives. And finally, the archive could store Webcam surveillance images or filling-level information, compare the data with target values using intelligent evaluation mechanisms in the network, and transmit any necessary alarm messages to the desired output devices.

Television/User-Generated Content

Triple play and mobile TV provide us with familiar television content using a different medium or an additional output device. The integration of additional information that users can retrieve depending on the context would constitute a usage-oriented approach. The convergence of primary content, supplementary content, comments, image material, and assessments ongoing on the Internet tends to interest people with Web affinity. That makes it all the more necessary for providers to gradually lead more hesitant customer groups to the future networked world on the back of the convergence of television and

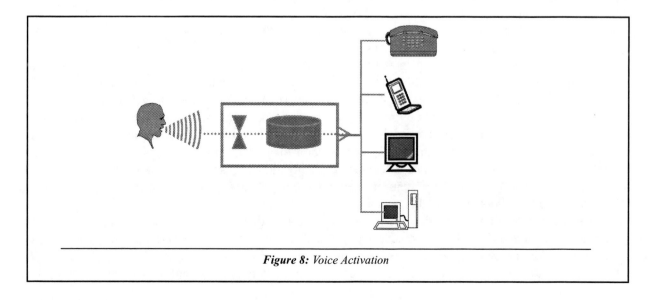

Figure 8: Voice Activation

Internet. Providers could likewise take the theme of user-generated content to these customer groups, without actually bouncing right into the blogosphere occupied by Web ultra-enthusiasts. A lot of people are organized in clubs and associations, all of which rely on exchange, information, and interaction. It would make sense to provide tools that would allow, say, amateur photographers or filmmakers to store the material taken at Sunday's game on a virtual archive, where they could edit it, supplement it with additional club material or material contributed by others, and make the result directly available on-line. A printing service could be additionally offered, for example, to make the club magazine available in hard-copy form for "traditional" members. Supplementary offerings for targeted club advertising and the incorporation of commercial links would allow additional sources of income to be tapped.

Applications/Connected Living

The location focus of our lives is strongly centered on our home location. We are all somehow at home somewhere—some people in a more stationary way, others more nomadically. What we all have in common is a need to control and organize that respective environment. For years now, we have seen a variety of strategies for networked living and smart homes. These have been made difficult by insufficient technology and prices that do not cater to the mass market. The situation changed with the advent of wireless technologies such as WLAN, Bluetooth, near-field communication (NFC), and universal plug and play (UPNP); the miniaturization of devices; and utility convergence in the home environment. The remote surveillance of heating systems, electrical appliances, or windows and doors is now possible. To optimize personal workflows, subsequent development steps will enable the use of things such as an interactive shopping list. We are not talking about the proverbial refrigerator that puts out an order for yogurt when supplies are running low. What we do mean is a direct setting on the PC/TV/family whiteboard for keeping an updated network-centric list of items needed, with that list made universally available via all sorts of devices and applications, as well as the possible integration of a payment function (contactless payment). It would be possible to further develop such a feature using location services aligned to the geo-position of the group/family members who would transmit the relevant order to the individual whose daily travel route takes him or her within proximity of the relevant shops. A direct 3-D navigation help could be used to provide information on routes and shop addresses as well as push information on special offerings.

Personal Service/Personal Care

A growing population of singles and a demographic shift in society are continuing to dissolve communities. There is also demand being generated for individual offerings for personal fitness and health (summarized by the term "personal care"). Newer mobile devices combine sensor functions, communications functions, and possibly localization (global positioning system [GPS] cell phones, ski-pass watches, radio frequency identification [RFID]). These are part of machine-to-machine communication. We can capture fitness data while running or at the gym, store it in virtual archives, subject it to a comparison of target and actual data, generate improved training plans, or receive personal training advice. Virtual training courses presented on TV or personalized exercise regimes using game consoles help to add more convenience to our individual lives. Where people with greater health needs are concerned, it is possible to use services that directly transmit data to health care specialists so that these may provide help on the phone, television, or another medium. Combined with geo-localization, offerings of this kind can help increase the mobility radius for these people, since qualified emergency help is always directly at hand.

Conclusion

Allow me to refer back to my opening words: This essay fervently makes a twofold appeal. On one hand, it aims to encourage those involved about the inherent potential in the services yet to be developed in the convergence area. On the other hand, it aims to emphasize the merits of re-engineering productization by reshifting the focus from technology to genuine customer segmentation.

Offerings that remain focused on a singular strategy will not reach users lastingly and will thus exploit neither business nor societal potential. The offerings we crudely refer to as multi-play are not sufficiently

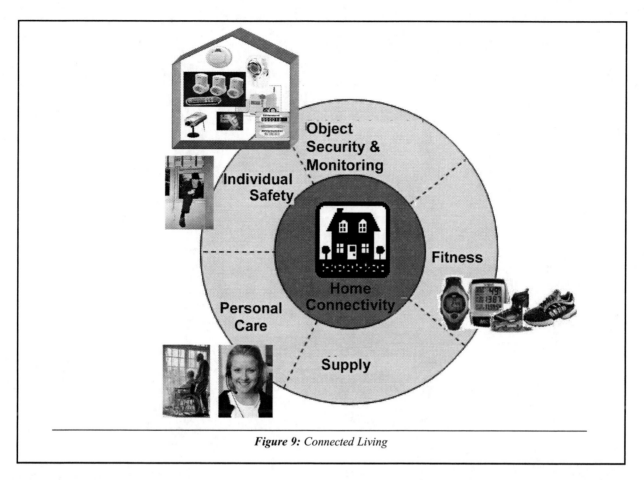

Figure 9: Connected Living

persuasive. The customer-centric and integrated ecosystem approach outlined above can represent an alternative bridgehead along the way.

Service generation and even marketing will continue to become more complex. Most of us are well aware of the technical component, even though it remains unmastered. The necessity to transcend current sector boundaries and create end-to-end business models is a far more demanding requirement. The transformation of the media industry clearly reflects the redesign requirements in terms of integrated content offerings, end-to-end personal care services, and secure transaction services.

Until now, we as an industry have remained rather unresponsive, both to our customers and to the idea of relevant value-creation partners. Alternative providers have unexpectedly moved to the fast lane. And just as unexpectedly, customers are becoming emancipated (right now, mainly on the Internet via Web x.0) and are surprising us.

The appeal I make is that we should open up our minds and give open thought to our actions so that we may turn our promises of innovative options into sustainable reality.

Fixed-Mobile Convergence – A Step toward Quadruple Play

Ramgopal Rajan
Senior Software Engineer
Wipro Technologies

Fixed-Mobile Convergence

Advances in telecommunications have made information sharing a global phenomenon. However, traditionally, most major technological breakthroughs have been in fixed-line implementations, making accessibility heavily location-dependent. For users to make and receive voice calls, they had to be at their homes or offices where they had access to a fixed-line telephone. To access data or watch a video, they had to be close to a computer or television. With the advent of pagers, mobile phones, laptops, and supporting technologies such as Global System for Mobile Communications (GSM), code division multiple access (CDMA), 802.11x, etc., the location dependency has been virtually wiped out, at least for voice calls and basic data exchange. With these changes in information accessibility, the general lifestyles of people too have changed. Even when on the move, the modern employee has alternatives to carry out various personal and professional commitments in an efficient, well-planned manner. Workforce accessibility, even during the after-office hours, gives the enterprises better resource utilization opportunities, which in turn gives them an edge in their business. However, with the concept of the mobile employee becoming commonplace, it is no longer a dominant differentiating factor in the marketplace. Enterprises are now looking at enhancing reachability with profitability margin consolidations.

The present-day wireless local-area network (WLAN) services are not robust enough to sustain the huge demands of the highly competitive and quickly expanding corporate world. The existing mobile technologies come at a premium, which stings the company finances. Even with increased spending, the enterprises are not able to exploit the mobility to its complete extent. Employees have to deal with multiple phone numbers, voice mails, text messages, etc., which affect their effectiveness and productivity. In today's world, mobility comes at a price, in terms of both resources and finances.

In such an environment, the concept of fixed-mobile convergence (FMC)—the integration of wireline and wireless solutions to create a single telecommunications network backbone—is a big plus for the service providers and enterprises alike. The convergence of fixed and mobile networks means that service providers can provide services to users irrespective of their location, access technology, and terminal, thereby bringing all potential customers within their realm.

This paper looks at the voice-based FMC solutions being rolled out by vendors in collaboration with mobile handset and fixed-line phone makers to their enterprise customers; the concept of extension to cellular. It then projects the extension of FMC solutions to non-enterprise users and to other services, namely data and video. The paper finally addresses the importance of Internet protocol multimedia subsystem (IMS) in achieving ubiquitous, seamless accessibility to all kinds of services.

Evolution of FMC Implementation

Nearly a quarter of a century ago, the concept of mobility got its first boost. The early forms of pagers and mobiles, though bulky and difficult to handle got instant acceptance from the business class users. The enterprise users latched on to their new-found freedom of mobility. With this early boost, the wireless industry took off and eventually spread into mass consumer space. Since then competing technologies have evolved, trying to get the larger pie of the huge mobile market world wide. GSM is the most popular solution available today in the mobile space. However, CDMA technologies with better bandwidth utilization and allocation present a stiff challenge to GSM. To gain an upper hand in this clash, the third-generation partnership project (3GPP), the consortium of global GSM players, came up with solutions such as general packet radio service (GPRS) and enhance data rates for GSM evolution (EDGE) to compete with CDMA2000 and wideband CDMA (WCDMA) solutions.

An offshoot of this continual tug-of-war between mobile technologies is the concept of convergence. The early form of FMC was the use of unlicensed mobile access (UMA). UMA provides GSM services over WLANs and/or ad hoc radio networks (e.g., Bluetooth) with built-in roaming and seamless handover between the two. However, being GSM–specific, it is not seen by many as the ideal solution for the future. One that fits this bill is the voice call continuity (VCC) technology. VCC extends an IMS network to cellular coverage and addresses handover.

Unlicensed Mobile Access

UMA was one of the pioneering technologies to support the concept of convergence. Though it does not support FMC, per se, it does lay the basic foundation: the use of dual-mode phones to achieve it.

UMA uses unlicensed spectrum technologies such as Bluetooth and Wi-Fi to achieve convergence. It is not unusual for enterprises to have their own ad hoc or Wi-Fi networks for internal communications. UMA uses these existing infrastructures as an alternative to the cellular radio access networks (RANs) to transmit the information to the mobile switching center (MSC). The UMA controllers ensure seamless coverage and handovers between the enterprise wireless networks and the cellular networks.

However, UMA basically aims at providing convergence on a GSM–specific environment with the primary aim of compatibility with the legacy GSM/GPRS networks. It changes the carrier at the lower layer and transmits the voice and data signal. The call information transfer retains all other characteristics of a regular mobile call (voice call or data transfer). In the unlicensed networks, the IP backbone is used to transfer the data only to the MSC; effectively, UMA is not an end-to-end voice over IP (VoIP) solution. Consequently, it does not have built-in provision for multi-person calls, instant messaging features, presence information, etc. It does not leverage SIP–compliant terminals, which are likely to be implemented on all WLAN–compatible terminals in the long term.

IP Multimedia Subsystem – Voice Call Continuity

IMS–VCC was the result of a school of thought that true mobility should encompass all possible market segments and should not be technology specific. It is a session initiation protocol (SIP)–based solution to the convergence conundrum. Being IP–based, it supports several enterprise-level call capabilities such as conference and transfer. It provides seamless VCC between the cellular domain and any IP–connectivity access networks that support VoIP. It is the most comprehensive of converged service approaches in that it can work between any cellular technology (e.g., GSM, UMTS, CDMA) and any VoIP–capable wireless access. IMS–VCC provides for the use of a single phone number (or SIP identity) as well as handover between WLAN and cellular.

Extension to Cellular – The First Step toward FMC

The FMC solutions aim at moving the information accessibility from fixed-line services to wireless ones. This is, generally, viewed as a major threat to fixed operators and a major boost to cellular operators. However, if one looks at the FMC concept, it is the ultimate realization that wired and wireless services have to co-exist in the communication world of tomorrow and there is not other alternative to this fact. What FMC ensures is that both can be used interchangeably, and more important, seamlessly. The first signs of the corporate world experimenting with the concept of FMC are apparent in the deployment of extension to cellular (EC) application that allows enterprises to connect their IP telephony platform to the cellular network. Given below are the two types of handsets on which EC solutions are currently employed in the corporate world:

FMC–enabled Handsets
Single Mode
The single-mode phones form an ideal platform to introduce FMC in its modest form, EC. The single-mode phones support a single mode of operation, i.e., GSM/CDMA. The EC solutions involve tweaking the enterprise switch to give the employee a convergence experience. The enterprise system administrator maps the employee desktop extension (physical or virtual) with the employee mobile phone number on the switch. Whenever a call is made from the employee phone, the user has to first call the switch and from then on the switch simulates the call on the mapped desktop phone, making access to inter-office extensions simpler and to outside calls cheaper. For any call on the employee extension, the switch places a simultaneous call to the mobile phone. Effectively, the employee receives the call on the desktop phone as well as mobile phone. Depending on the location, the employee can answer the call on the mobile or desktop phone.

Dual Mode
The dual-mode phones provide the true FMC that the telcos dream about. Dual-mode handsets are the ones that support two technologies, usually GSM/CDMA and Wi-Fi. In this case, the GSM ID (cell phone number) and the Wi-Fi ID (usually something like the SIP profile) of the dual-mode phone are mapped to the employee extension. The Wi-Fi support extends the internal call availability right up to the range of an enterprise Wi-Fi network. The phone can automatically drop the GSM call and connect to a Wi-Fi call as soon as one enters the company campus. This handover happens seamlessly and is transparent to the caller at the other end. With automatic handoffs, dual-mode handsets enhance user EC experience and ensure more efficient call handling at the same time.

Enterprise FMC Deployment

When the mobile revolution first began, it was the corporate world that embraced it with both hands, giving it the fillip to become a worldwide phenomenon. Now, we are at the threshold of another telecom storm, the convergence wave. The telcos again are looking up the enterprises to provide a foundation to push FMC and quadruple play forward. The organized customer base and a central or well-connected network of internal switches in the enterprises makes them ideal platforms to launch the FMC services. Enterprise segments provide the operators the structured basic landline and mobile infrastructure required for convergence. Besides, they also provide ample opportunities to scale the FMC deployment. With a strong enterprise presence, operators can roll out services on a mass scale too. The corporate experience would augur well for the operators in this stead.

While the telcos look at corporate houses for support in this regard, it is not a one-way, philanthropic relationship. In turn the companies also benefit. The following are some of the immediate advantages that the enterprise will have, giving it an edge in the market:

- Single contact number/work in office from anywhere—FMC solutions provide virtual extensions on employees' mobile phones. With FMC, irrespective of the current location, users can always be reached on their office extensions.
- Reduced costs for international calls—This is probably one of the biggest economic benefits of deploying enterprise FMC solutions. Any official call that employees need to make when on the move can be routed through their desktop phone through the company's internal telecom infrastructure. With such a solution, the enterprise would see a huge savings on the official GSM mobile bills.
- Seamless roaming—With dual mode, the advantage of seamless roaming can be fully exploited. When out of the local office Wi-Fi area, the user would be accessible through the GSM network. In this scenario, the dual-mode phone becomes a virtual desktop phone. Once within the Wi-Fi network, the phone/software

running on the phone can automatically/manually switch to the Wi-Fi coverage and drop the GSM call. This kind of seamless roaming helps use resources effectively without affecting the employee or client experience.

- Continuously reachable workforce—In today's fast-evolving global business world, a 24-hour reachable workforce can give companies a competitive edge over their peers. With FMC solutions, employees can be contacted on a single number, making accessibility simpler for clients and the enterprise alike. FMC solutions can make productivity transparent of work location.

Extension to Cellular Implementation for Internal Calls

The overnight software build failed due to some technical glitch in the code. Initial troubleshooting has shown that the code checked in by Ram has caused the build to fail. Harry, the build manager, calls Ram's extension to talk with him, unaware of

the fact that Ram is stuck in traffic. With the EC solutions implemented in the enterprise, Harry's call gets extended to Ram's dual-mode phone. Ram picks the incoming GSM call. While the call is in progress, Ram reaches his Wi-Fi–enabled office campus. As soon as the dual-mode phone senses the Wi-Fi network, it registers itself in the network, sets an SIP call with the enterprise switch, and drops the GSM call. With this switch, the GSM call (between the enterprise switch and Ram's mobile phone) ends and now the call is an internal call being routed through the enterprise switch and SIP server. When Ram reaches his desk, he bridges his desktop IP phone on to the call and drops the Wi-Fi call. All the handovers and switches are transparent to Harry.

Extension to Cellular Implementation for International Outgoing Calls

Mark has to make an important international call to one of his clients. He has been held up because of equally important work at his house. Mark places

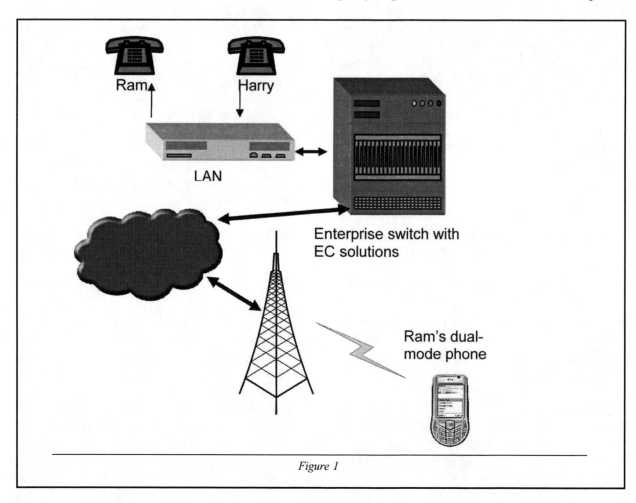

Ram

Harry

LAN

Enterprise switch with EC solutions

Ram's dual-mode phone

Figure 1

the call using his dual-mode mobile phone. The call first goes to the local enterprise switch as a GSM call. The switch then makes an international call on the company's backbone network. While the call is active, Mark finishes his household commitments and makes it to the office. Here, the handset switches to the Wi-Fi network in the office and ends the GSM call. Mark can now reach his desk and extend the call to his desktop phone if he wishes to. The entire process of handover and switching is transparent to Mark's customer.

In both cases, the enterprise makes savings in premature termination of the GSM call, without affecting the chat between the parties. The international calls being routed through the enterprise switch leads to further savings. Besides these tangible advantages, with EC solutions, the enterprise also makes huge intangible savings. By providing location-independent access to its employees, it has ensured an all-important professional experience for its client.

Importance of IMS

All through the paper we have been looking at FMC solutions for enterprises. On similar lines, though, the concept of FMC can be extended to the non-enterprise users too. However, with retail users it is not a single enterprise that integrates them; it is the carrier or the service provider. Currently, service providers do not have a system wherein the user landline phones are mapped to the user mobile phones. With such a mapping in place, the service providers would be able to unify the fixed landline and mobile GSM/CDMA solutions, providing true FMC solutions to the public. Besides, voice calls to provide a comprehensive convergence experience to the end user, telcos, and vendors are looking at a quadruple-play deployment: a telecom environment where the triple-play services (e.g., bundled voice, video, data) can be ported on to the mobile devices.

Most carriers would use different approaches to provide triple-play services. Coaxial cable, digital subscriber line (DSL), and passive optical network (PON) are some of the most common ones. Even on the wireless front the eternal competition between GSM and CDMA would continue. This would mean that in a quadruple-play market, customers

would have to make a lot more initial investment to switch from one carrier to another. On the other hand, converged service access would mean the expenditure would go southward, making the package very attractive. Reduced bills and costly service provider swaps would improve customer loyalty. Besides customer loyalty, reduced maintenance and distribution costs for quadruple-play deployment makes it a lucrative option for service providers too.

For such ambitious solutions, current telecom fraternity is looking at IMS to do the trick. IMS is the universal IP–based backbone that would present itself in the front end in the form of existing fixed-line and wireless technologies. In an IMS world, the Wi-Fi access point, the Class V switch, the GSM base transceiver station (BTS), etc., all get connected to the backbone with specific interfaces. With such a universal interconnection, the IMS can keep track of every call and information exchange that takes place, and as the user moves across various media, the IMS can effectively make new connections and drop older ones in a manner transparent to the user.

IMS is envisioned to provide end users with ubiquitous access to the triple-play services on offer. For this to be implemented successfully, it is imperative to have a common unified database of all the access modes that a user has registered with and an ability to link them when the user intends to switch modes. The biggest hurdle today is that most service providers have the capabilities to provide bundled triple-play services but not all the modes of access that a user may prefer to use. In this regard, service providers would have to share their customer information on a common database, which is not something that they would be open to. To most, it would be losing out on the customer, as this would generate a different kind of customer churn. Though customers are registered with them and use their services loyally, they may not use them exclusively, and this would affect service providers.

Conclusion

The deployment of triple-play services was a new wave of telecom revolution in mature markets. The concept of delivering voice, video, and data services in a unified manner was the way forward for most incumbent local-exchange carriers (ILECs) to ensure better profitability and customer loyalty. Since its inception, mobile communications has been growing at a very fast rate, creating and expanding markets worldwide. With advances in triple-play and wireless technologies and markets, the confluence of fixed and mobile technologies has become viable—if not immediately, then at least in the foreseeable future. The corporate world again is leading the way to embrace this new kid on the block. Most product and equipment vendors are working on developing solutions for FMC. Early deployments on EC have already been a success and have received a good response in the enterprise world. With improved employee reachability and productivity, reduced telecom costs, streamlined management, and lowered operations expenses, enterprises are finding a good ally in the FMC solutions to up their profits.

Today's business workers have many options for communicating, ranging from the traditional paper mail, fax, voice (landline and mobile), and e-mail to the latest conferencing (voice and video) and instant messaging. This has given them several avenues to ensure improved reachability and response time. Unfortunately, this diverse range of communications and messaging modes have actually degraded their responsiveness, as workers end up spending too much time managing their diversified accounts. This is having a negative effect, as they get distracted from the critical business activities that they need to address. The evolution of FMC is seen as a transition from service convergence to identity convergence where the user can be universally contacted through a single number.

FMC solutions form an ideal steppingstone for the eventual deployment of quadruple-play Services. In this endeavor, IMS would play an important part. Service providers who can provide wireline and wireless services to their customers can look at incorporating the IMS backbone into their deployment. With IMS in place, the carriers can link the mobile and fixed-line solutions provided to their customers. With this unified offering, customer churn can be greatly reduced, as all the services are offered in a bundled form.

It is a well-known fact that fixed-line services are here to stay. Considering their usage, their penetration, and the finances and resources invested, it would be extremely difficult for service providers to extirpate them completely. However, wireless solutions would provide the service providers an opportunity to improve the realm of service being provided. For the user, wireless would give them an additional option to be in touch even when they are on the move. In such a converged environment, with both fixed-line and wireless solutions complementing each other, IMS is envisioned to play a very important role.

Beyond Triple Play

In Quadruple Play, Delivering Exceptional Customer Service Is More Critical (and Difficult) than Ever

Michael Roy

Vice President, Customer Management
StarTek, Inc.

While the success of triple play—video, high-speed Internet/data, and premises-based voice convergence—has been making headlines, it is the quadruple play and quad-play plus that are promising to move the needle even more dramatically for increased customer value, retention, and improved revenue streams. By adding mobility and focused content to the mix of the three current services, service providers have the compelling ingredient that will lure yet more customers away from single or dual services to a bundled solution of three, four, or more integrated offerings.

Now for the bad news: competition is fierce for the higher-value, bundled services customer and it is coming from everywhere—traditional providers and many new competitive sources. With bundled services, technological innovations, and industry deregulation, the traditional lines of demarcation are being moved or eliminated, resulting in mergers, consolidations, partnerships, and other business model–shifting trends.

The deciding factor that determines which companies will succeed will ultimately come down to customers' brand experience, the majority of which is formed through customer support interactions. This article focuses on what service providers should know about the unique support challenges companies face in offering triple-/quadruple-service bundles and how they need to prepare to ensure the highest level of contact center service to these valuable customers.

Quad Play Takes Off with Consumers ...

There is no question that the quadruple play—with the addition of mobility—is going to be a boon for service providers and consumers alike. For customers, it is one-stop shopping, billing simplicity, and the prospect of a lower total monthly outlay. For service providers, it offers increased customer loyalty, competitive advantage, and a higher value set of customers.

This win-win proposition is gaining steam in the market. According to research from Parks Associates, by 2010 more than half of broadband households will subscribe to a triple- or quad-play bundled service. IDC is forecasting a significant drop-off of double-play service over the next five years, falling to 42 percent of total subscriptions by 2010.

... While Service Providers Enjoy Increased Stickiness

Service providers already offering triple- and quad-play service offerings are beginning to reap the financial benefits of the multi-play strategy. One area where companies are seeing dramatic results is customer "stickiness," or loyalty. According to Frost and Sullivan/Stratecast, in the North American cable market, a three-product subscriber is as much as 50 to 60 percent less likely to churn than a single-product subscriber.

However, Success Requires Major Operational Changes

Whether through partnerships, acquisitions, or creating new business models, delivering a successful triple- or quad-play bundle will require operational changes and strategies that ensure a smooth rollout and seamless customer experience throughout. New higher-value customers represent not only increased revenues, but increased risk and expense; particularly in cases of customer churn. Therefore, getting the service delivery, operational, and customer care aspects right is essential to ensure the excellent brand experience that will separate the market winners from the losers.

With technology hurdles and rapid introduction of new and changing services, the onus will fall to customer support to ensure customer satisfaction while hiding the complexity of delivering disparate services from the customer. In this high-stakes environment, customer support will need to surmount several major hurdles, including the following, to deliver a brand experience that exceeds expectations:

- Uneven quality of services
- Integration of multiple systems and products for one bill/one voice
- Overcoming technological complexity

Quality of Service Is Only as Strong as the Weakest Link

Customers purchasing bundled services will expect a high level of support as well as consistent, high-quality experiences. In fact, each service must be uniformly good quality or it jeopardizes the entire bundle. Unfortunately, providing all the services within a bundle can be a disjointed process given the fact that products may originate from a variety of sources thanks to inherent technologies, subcontracts, joint ventures, acquired companies, and other business models. Customer support must develop ways to shield the customer from these issues, yet deliver a coordinated view of the subscriber and a well-orchestrated experience to questions, billing, and service problems.

One Bill/One Voice – Fulfilling the One-Stop Shopping Expectation

While quad play means one stop to customers, in reality many vendors have multiple systems across multiple service offerings, making customer care anything but seamless. Regardless of where the service is ultimately originating, the customer will expect one company to support all the services provided. Care agents, business processes, and support systems will be required to integrate and align as never before.

Technological Complexity – Shielding the Customer

Customer care agents will be expected to provide support and resolutions at a higher level, across a larger matrix of products and services. Rather than the tactical challenges of a phone connection being down, cable TV being out, or digital subscriber line (DSL) not working, multi-play support issues can present any combination of these issues at varying levels of complexity. Those providers that can keep the frustration of complex technology delivery invisible to the end user will be the ones most likely to succeed.

According to IDC analyst Brian Bingham, global director of CRM Services and F&A BPO Research, the role of customer service representative (CSR) has evolved, especially in the communications/cable industries.

"Not only does the CSR need to understand the basic voice service offerings," Mr. Bingham said, "but now with the introduction of quad-plus-play services (voice, video, broadband, mobile, and media/content), they need to troubleshoot complex service outage questions, understand ever-changing regulations, and a whole host of other technical processes and systems. That is why we are seeing an uptick of traditional wireline, wireless, and major cable companies gravitating toward outsourcing providers to deliver bottom-line savings."

Addressing the Customer Service Challenges of Multi-Play

The challenges of quad play require service providers to work toward customer support solutions that provide high levels of first-call resolution, care agents who are skilled across multiple offerings, and agents who can provide information about and processing of additional, value-added services. At the same time, intense competition for wallet share means that service providers must also ensure a cost-effective infrastructure to protect their profit margins while growing their customer base.

Companies can choose to handle the customer care function in-house, addressing the above-mentioned challenges of complexity, one voice, and quality of service through increased staffing, extensive agent training, new systems, and updated technology. Cost control can then become an issue and often results in either less-than-adequate staffing for call volume peaks or overcapacity, which translates into cost overruns. When the center is understaffed, call abandonment rates and call wait times escalate, payroll costs increase due to overtime hours, agents experience fatigue and burnout, processes break down, and customer satisfaction declines.

The Value of Outsourcing Some or All of Multi-Play Customer Support

When faced with the choice between internal or outsourced fulfillment of the complex customer care requirements of triple- or quad-plus-play strategies, many companies are seeing the value of identifying and sourcing services and strategy with a trusted partner.

A proven outsourcer's standardized processes and flexible staffing allow it to leverage experience, infrastructure, and capacity, eliminating the service provider's need to overinvest in resources. The right outsourcing partner can offer experienced management, well-trained agents, and systems for the varied types of services being offered in the new bundles, ensuring superior first-call resolution and customer satisfaction.

However, companies leveraging outsourcing as a cost-effective solution to the support challenges of multi-play strategies need to research and hire a business and brand partner, not merely an outsource services provider. Organizations should seek out a collaborative partner with complementary resources and technology and a focused common interest in not just handling a subscriber event, but also meeting the strategic needs of the company.

You Have to Play to Win

Seen as a win-win for both consumers and service providers, the quad-play strategy also brings a number of challenges with it: new and upgraded technology, disparate services and sources of services, complexity of interrelated services, and integrating business processes to provide one-stop shopping and support to the customer. Managing the total customer experience throughout the customer life cycle will be critical to success. Quality customer care becomes the lynchpin element to ensure a seamless and satisfying subscriber experience with the multi-play bundle.

IMS – The Ideal Architecture for Enabling Quadruple Play for Operators

Arjun Roychowdhury

Director, Converged Applications and IMS
Hughes Systique Corporation

Abstract

Broadband Internet protocol (IP) is a great leveling ground when it comes to converged services being offered by multiple providers. For example, with the availability of broadband, companies such as Vonage are offering IP–based phone replacement solutions, threatening the turf of established phone operators. Similarly, Comcast can now offer cable voice over IP (VoIP) and Verizon can now offer TV services over IP, thereby threatening other companies specializing in service areas that were traditionally never their turf. Broadband IP has also enabled "new kids on the block" such as Skype and Joost to offer bundled services that threaten the trillion-dollar communications industry as we know it. This is one main reason why carriers are competing to stay alive with quadruple-play blended services that offer voice, video, data, and wireless accessibility into one.

However, providing quadruple play across heterogeneous networks (e.g., worldwide interoperability for microwave access [WiMAX], digital subscriber line [DSL], cable, cellular) is a non-trivial task, and one needs a robust and well-thought-out architecture that ensures services can be provisioned and provided uniformly to subscribers in a way that lends to seamless user experience and operator provisioning/charging and billing.

This paper describes the merit of the IP multimedia subsystem (IMS)—an overarching architecture specification that enables uniform IP–based service delivery over diverse network types (e.g., wireless fidelity [Wi-Fi], DSL, WiMAX, cellular technologies) as the ideal architecture for operators to deliver quad-play services to their users.

Uniform Service Delivery – A Key Requirement for Quadruple Play

Uniform service delivery is a key requirement for delivering effective quadruple-play services.

For example, how does Bob, who is watching an IPTV stream on his TV, walk out of his office and continue to receive the IPTV stream (with possibly reduced quality) on his high-speed downlink packet access (HSDPA) cell phone? Alternately, Bob, while talking to Alice on his phone, should be able to hand over his call to his desktop computer over DSL so that the existing call could be enhanced with on-line collaboration or photo sharing (assuming Alice's device supports it).

Providing a uniform service delivery architecture can be translated into the following key requirements:
- Ensure subscribers and devices can be identified uniquely for the purposes of charging and delivery of services across different access mediums
- Ensure security, integrity, and authentication is maintained
- Ensure a common policy (quality of service [QoS]) and charging architecture
- Provide voice call continuity (often referred to as basic service continuity)

The layered third-generation partnership protocol (3GPP)/IMS architecture attempts to enable all of the above. This paper will discuss in more detail how IMS enables the above and therefore is ideally suited to a heterogeneous network.

The rest of the document will provide more insight into how IMS enables the four key elements sighted above, further justifying our belief that IMS, as it stands today, is the most apt architecture for heterogeneous network deployment.

Benefits of a Uniform Service Delivery Architecture

The benefits of a uniform service delivery architecture are as follows:

- Reduced migration cost—Legacy systems are plagued with the cost of service re-investment. For example, today, to migrate an existing DSL subscriber service to a Wi-Fi service requires significant rework for back-end provisioning systems, since the data formats are not abstracted sufficiently and the service provisioning makes direct assumptions about the access infrastructure. The IMS enabled concept of home subscriber server (HSS) data abstraction and global user profile (GUP) reduces this significantly.
- Reduced operational service deployment cost—Any service that has been deployed at the IMS level—for example, call transfer—will work the same way whether the subscriber is on a general packet radio service (GPRS) network or a DSL network. This is because the IMS service execution architecture is sufficiently abstracted from the access network details and any access-level changes only affect the respective layer nodes without affecting higher-level nodes in the 3G layered architecture (a classic layered network approach).
- Increased vendor interoperability— IMS–defined explicit interfaces between the core network (e.g., call session control function [CSCF]) and the application servers, which when adhered to makes it significantly simpler for third-party application servers to participate in an operator network to enhance value-added services to the subscriber. In fact, fortunately, in IMS, the IMS service control (ISC) interface

(which is session initiation protocol [SIP]) between the core IMS network and the application server is the least complex of all other interfaces, and the authors have worked with several independent software vendors (ISVs) that have deployed services in non–IMS networks and have easily managed to deploy the same services with cellular networks, using IMS, by adding incremental support for Sh (Diameter for HSS), SIP header extensions, and compliance to the IMS generic charging model.

- Reasonably future-proof—Since the IMS architecture has already defined (or is in the process of defining) interworking profiles for a wide variety of existing access technologies and has a generic architecture that should be able to accommodate new access technologies, this gives the architecture a reasonable insurance against disruptive network topologies that may evolve in the future (we say reasonable because future-proofing is always a best-effort activity based on current visibility of what the future would be like).

Enabling Quadruple Play via a Uniform Service Delivery Platform

This section details how the IMS architecture provides an ideal platform for enabling quadruple play.

Subscriber Identity

IMS has specified a generic subscriber identity mechanism that is independent of the underlying network. In addition, it is very easy to associate more than one service profile to a single subscriber. This is best described by the following example:

Let us suppose that Joe is an IMS subscriber. While he uses a code division multiple access (CDMA) network to connect to the IMS services, due to the limited bandwidth of the network, he is only allowed to access voice calls at a 5.3 kbps codec and basic e-mail. However, when Joe connects via a high-speed WiMAX network, he is allowed to use a 64 kbps voice codec, make video calls, and access streaming IPTV on his WiMAX device. This inherently means the following:

- The subscriber identification should be generic so that it can be used across any access network.

- The network needs to correlate all these scenarios to a single user (Joe).
- The network needs to provide for the fact that Joe may have one or more service profiles active at any time (for example, Joe may be registered simultaneously with both his WiMAX personal digital assistant [PDA] and CDMA cell phone)
- It is also possible that Joe may use multiple devices, so subscriber-to-device mapping should not affect operation.

IMS solves this by introducing the concept of public and private uniform resource identifiers (URIs) and service profiles.

All identities in IMS are standardized as SIP URIs. Using the SIP URI addressing scheme ensures that identification is independent of the underlying network. If any mapping is needed, that happens depending on the network type attached to. For example, when connected to the Universal Mobile Telecommunications System (UMTS), the URI may be mapped to a mobile station integrated services digital network (MSISDN) or International Mobile Subscriber Identification (IMSI) (for public and private URI).

In IMS, the public URI is the "public" address through which the entity can be addressed. The private URI is a network- or operator-assigned URI that ensures that the user is correctly authenticated and verified by the network (more on this later). A single private user identity can have more than one public user identity.

Finally, each public URI can be linked to a service profile that describes the available services that are active for that profile. It is also possible that a single service profile is shared between more than one public URI.

Diagrammatically, the relationship between subscription, private, public, and service profile is as shown in *Figure 1* (source TS 23.228).

Security: Authentication, Confidentiality, Integrity

One of the biggest challenges in being able to provide uniform services across networks is to ensure that security is maintained. For a closed network configuration, ensuring security is a simpler task, since all layers are known. However, since IMS can run over multiple access networks, it is possible that some access networks provide strong security while others do not. Therefore IMS provides its own security negotiation at the IMS level, irrespective of underlying security negotiations that may have already taken place. For example, when Wi-Fi user equipment (UE) is first turned on, it may have already established an encrypted and secure connection with the Wi-Fi base station. However, when that UE "registers" to IMS, it re-establishes a new set of security parameters that is visible at the IMS layer. While some may think this is duplication of security, there are several reasons why this is done, including the following:

- As described previously, IMS may work over any access network type. It is possible that some access networks have weak security or no security at all. Not doing IMS–level security

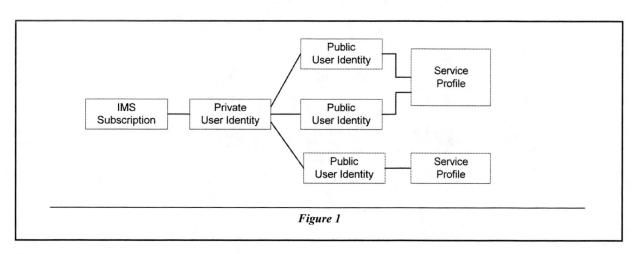

Figure 1

compromises the IMS nodes as well as the subscriber profile.

- It is likely that the access network may have been "hijacked"—rogue UE could impersonate another access network device if the security is not strong. Therefore, the IMS cannot assume that a verified access identity is a verified service identity.

IMS provides security, authentication, and integrity in several ways, including the following:

- The UE connection to the proxy CSCF (P–CSCF) over-the-air interface (if applicable) is typically a two-way IP security association (SA). This enables the network to verify the user and the user to verify the network as well (what if a rogue "network" positions itself as the real network for a valid user to connect to?).
- Each subscriber in the IMS network is assigned a private URI. This private URI is typically stored in the ISIM of the mobile station and is not open for public view. A rogue client will need to access the ISIM to impersonate the user. (Note however, that some devices may not have an ISIM. This is typically the case when a DSL subscriber attempts to connect to IMS. In this case, alternate security mechanisms or a

soft–ISIM—a universal serial bus [USB] stick that provides the functionality of the ISIM—may be used).

As a side note, several vendors have felt that completely independent security mechanisms at different layers, while complete does result in more cycles and delay in fast attach to networks. In situations where there is some control on the access network attachments, vendors have proposed mechanisms where security from a sub-layer is recognized and percolated at the IMS layer to reduce security negotiation delays.

Diagrammatically, the IMS security relationships are shown in *Figure 2* (source TS 33.203).

In *Figure 2*, (1) and (2) are the private key, mutual authentication, and IPSec SA, while (3) and (5) deal with intra-domain security (Cx interface) and (4) deals with inter-domain security.

As far as integrity and confidentiality protection goes, this needs to be applied to both UE–PCSCF as well as CSCF–CSCF. The integrity protection between UE–PCSCF is based on IMS authentication and key agreement (AKA) (a derivative of the UMTS AKA specification). By this mechanism,

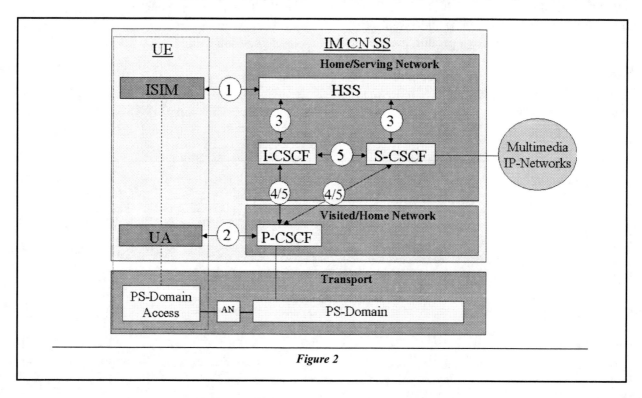

Figure 2

both the UE and the P–CSCF verify that the data sent from each other has not been tampered with. Typically the integrity protection between CSCFs is via transport layer security (TLS).

Common Policy and Charging Architecture

With subscriber identity and security out of the way, the next challenge comes in providing a common infrastructure where policies as well as charging information can be regulated irrespective of the access network. This particular problem is harder than it sounds. To give the reader an idea of its complexities, consider a case where two users are connected in a 128 kbps video call over a UMTS network. The gateway general support node (GGSN) has been instructed (i.e., policy) by the CSCF to allow 128 kbps of bandwidth for this particular call. Now the call hands over to a WiMAX network. This would mean that the GGSN is no longer the policy enforcement node. That role is now played by the access service network gateway (ASN–GW). How does the CSCF communicate with the ASN–GW to ensure that the same rules apply? Furthermore, how does one ensure that the charging information that was propagated by the GGSN is continued to be propagated by the ASN–GW? The problem is that while 3GPP specified specific rules, in Release 5,

about how the policy decision function (PDF) (usually the P–CSCF) controls the policy enforcement function (PEF) (usually the GGSN) in an UMTS network, there were no well-defined mechanisms to abstract this interface for other access networks. Specifically, WiMAX was specified by the Institute of Electrical and Electronics Engineering (IEEE), and the network architecture is being specified by the WiMAX forum, a separate body from the 3GPP. The Release 5 specific PDF–PEF functionality was not generic enough and did not yield to the WiMAX network being able to use all the potential advantages of a WiMAX network. This resulted in 3GPP starting an effort on a more generic policy and charging control (PCC) specification in 23.203.

The PCC functionalities are divided into two categories: charging control, which offers mechanisms to charge a session/stream based on on-line or off-line (used for services that are paid for periodically) mode, and policy control mechanisms that allow application of QoS and flow control to sessions/streams.

Figure 3 illustrates the reference architecture of PCC as defined by 3GPP. The policy and charging rules function (PCRF) is a logical entity that specifies

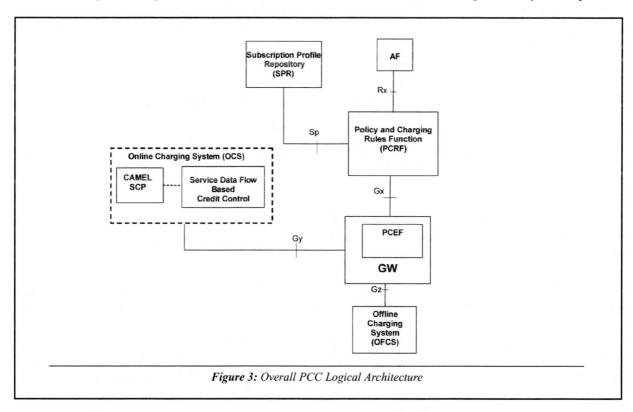

Figure 3: Overall PCC Logical Architecture

"rules" that define how policy and charging rules are applied in the network, and the policy control enforcement point (PCEF) has the role of receiving those rules and acting on them to enforce those rules.

Parameters for policy control and/or charging controls (based on detection of a service flow initiation to or from a mobile station [MS]) are collectively termed as a PCC rule. PCC rules are mostly IP connectivity access network (IP–CAN)–agnostic but have some IP–CAN–bearer-specific elements. An IP–CAN session incorporates one or multiple IP–CAN bearers (support for multiple IP–CAN bearers per IP–CAN session is IP–CAN–specific) and exists as long as a UE IP address is established.

The signaling between the PCRF and PCEF in a PCC framework (Gx reference point) is Diameter-based.

Since the PCRF and PCEF are "logical" nodes, in reality, they could be a part of a physical node. For example, the P–CSCF could behave like the PCRF and the GGSN like the PCEF. In addition, since the architecture is intended to be access-independent, in the case of WiMAX interworking with UMTS, the ASN–GW could be the PCEF in the WiMAX leg while the GGSN is the PCEF in the UMTS leg.

Voice Call Continuity

Voice call continuity (VCC) is an area of active standardization as of today. Currently in Stage 2 specifications (TS 23.206) at 3GPP, VCC provides an architecture that allows for origination, termination, and dynamic transfer of an SIP call over IMS to a circuit switched (CS) call and vice versa. Simply put, if Joe is talking to Bob over IMS and in conversation, and Bob roams out of IMS coverage to where only CS connections are possible (or the other way around), VCC allows for a handover in a way that neither Joe not Bob experiences a call disruption.

One of the key design requirements of VCC was to try and ensure that the core CSCF nodes are not loaded with the responsibilities of call continuity and ideally, this should be an application-level functionality that the CSCFs are transparent to. In an effort to meet the goal, the VCC server is essentially a SIP back-to-back user agent (B2BUA) that

serves as an anchor point for all calls for a particular VCC–enabled UE. This means, as follows:

- Any UE that requires CS/IMS call continuity needs to be aware of the VCC application server (VCC–AS) and registers with it at the start. As a corollary, the UE must be enhanced to support VCC interfaces to be able to participate.
- Any call that is made by that VCC UE is always anchored at the VCC server to ensure that proper handover is performed in-session, if required.
- At a high level, the VCC is an application server that resides behind the S–CSCF. The flow of normal signaling from UE to P–CSCF to I–CSCF to S–CSCF and then AS based on iFC is therefore preserved when it comes to IMS signaling. (Note that the VCC UE has a direct interface to the VCC–AS using the V3 reference point—this is typically realized using the Ut reference point and is related to management and provisioning of UE services, not specifically VCC–related. Examples of protocols that could be applied at the Ut reference point are HTTP and extensible markup language configuration access protocol (XCAP).

So from the IMS perspective, we understand that the VCC–AS is essentially a SIP–based B2BUA. But to be able to support CS calls, the VCC–AS also needs to support CS interfaces. The CS interworking is performed by components in the VCC–AS called the CS adaptation function (CSAF) and customized application of mobile enhanced logic (CAMEL) service. The scope of the protocol functionality of the CSAF and CAMEL services is too wide to be included in this paper—therefore, suffice to say, think of the CSAF and CAMEL service as providing the required functionality to route CS calls, maps CS identifiers such as the IP multimedia routing number (IMRN) to IMS routes, etc.

So now, we understand that the VCC–AS is a B2B entity that offers both SIP/IMS leg and CS leg support. We now need functionality that can switch between the two when required. The domain selection function (DSF) and the domain transfer function (DTF) provide this functionality—it provides interfaces and hooks to execute a call handover based on policy, provides current status of the VCC session (e.g., is the UE registered in IMS in the first place to be able to switch to an IMS session?) and more.

Finally, the last question is "When does the handover occur?" The answer to that is, it depends. VCC does not specify mandatory rules on detection of a handover. This is left to other protocols, either initiated by the UE or the network to decide when is the right time for a switch. Mechanisms include media-independent handover and link-layer signal assessment. Remember that the goal of VCC is to be able to work over as many access networks as possible, and it is likely that different access networks may have different detection and handover policies that are better suited for those networks.

Conclusion

In this paper, we have explored how IMS enables key requirements for a truly convergent network architecture and further justifies why we believe that as disruptive access technologies such as WiMAX are deployed, IMS will be required as a session control architecture to be able to deliver quadruple-play services across these diverse access technologies. When IMS was first introduced, it had several elements that were specific only to cellular networks and it was rightly felt that much of this may not be required in a non-cellular access stratum. Furthermore, there was no clear path of interworking with wireline deployments. However, as time evolved, the IMS architecture cleaned up significantly and dealt with issues of interworking with generic access stratums (wireline or wireless) as well as abstracted the architecture sufficiently to reduce dependency on specific radio access networks. Obviously, the work is not completely done yet, but it is our belief that IMS is the best architecture that is available today for architects to adapt to their networks. Starting from scratch would likely result in us having to reinvent a lot of what IMS has already done.

Quadruple-Play Services: A New Horizon in Telecom VAS

Rakesh Shukla

Managing Partner, Emerging Technology Markets
Amatra Business Consulting

Naveen Hegde

Research Associate, Emerging Technology Markets
Amatra Business Consulting

Introduction

The telecommunications industry is currently changing at a rapid pace, driven by the emergence of new technologies that are rewriting the business cases and cost models on which telephony has been based for years. The improved technology and the increased competition on account of globalization are in fact driving the new strategies of business in the telecom sector. As against the specialization of the businesses and the processes strategy used in manufacturing and other services, the telecom companies are in the lead to use packaging services strategies for efficient market capture in the near future. In fact the access network is being revolutionized by a combination of forces—demand for new broadband services such as fast Web downloads and Internet protocol television (IPTV); new competitive market entrants; regulatory regime change; and the advent of home networking, which means that a plethora of new devices is being connected to the telecom network through fixed and wireless connections.

In line with this, many telecom companies are collectively spending huge amounts to roll out the infrastructure and technologies that will drive the next generation of voice, data, and video services. These bundled services, known as the triple play, represent the basis for telecom product strategies over the next decade. Adding wireless to the package brings in a fourth dimension, thus the term quadruple play.

Triple Play Grows to Quad Play

In telecommunications, triple play is an IP service that brings together the Internet, television (video on demand [VoD] or live stream), and phone (voice over IP [VoIP]) services.

Triple play is also great for offering enhanced and converged services such as the following:

- TV and video services such as VoD, subscription, live TV channel lineup packaging and scheduling, and network-based digital video recorder capabilities
- Additional blended communications applications such as receiving phone call information on your TV, call reject or receive, or text chat
- Personal media applications such as uploading of photographs and video clips and sharing of music and video content across multiple in-home devices

This implies that users have a measure of simplicity and also (but not necessarily) that all services are over a single broadband IP connection. This does meet the test of perceived and real value, since studies do show significant reduction in customer churn when subscribers get double or triple play over one connection. Typically it comes with cost savings for the subscriber and some measure of simplicity for installation, use, and customer service. And there is also the benefit of having one bill.

Simple and accessible self-care and graphic user interfaces (GUIs)
- User-friendly subscription and service management
- Personal profile management, including advanced parental control
- TV-based and Web-based self-care
- High degree of customization and personalization options

Table 1

The access hub platform is the ideal choice for operators wishing to provide triple-play services via a single network technology. Despite the different nature of these services (IP–, asynchronous transfer mode [ATM]– and time division multiplexing [TDM]–based), access hub IP multiservices access network (MSAN) provides the necessary inter-working functions for them to be transported via the same IP–based backbone, avoiding duplicated core infrastructure.

The major milestone for any triple-play provider is delivering the services via a converged IP network. For many of them, this represents their first time building a converged network; it is a watershed of sorts for them as they realize the potential for service creation goes well beyond basic voice, video, and data services. In fact, once the IP network is in place (regardless of access technology), it becomes much easier to deploy other advanced IP services (e.g., gaming, presence applications, e-learning, e-medicine, music, collaboration, security, monitoring, meter reading, dating). All that is required is new service logic and, perhaps, some new customer-premises equipment (CPE) (e.g., video cameras for home monitoring).

The triple-play market for U.S. service providers is expected to grow from $137.5 billion in 2006 to $145.3 billion by 2009, according to a recently released study by the Yankee Group. Still, when it comes to triple-play services, the key business opportunity at hand is turning on the tens of millions of unconnected homes quickly with gateway-centric products that save time and money. This could represent a strategic advantage in this highly competitive market.

The "battle of the bundle" between the cable and telecom industries has heated up all summer as communities across the country are being besieged with discounted offers for triple-play services. And as both sides continue to chip away at each other's market, industry insiders say the price wars being waged will eventually lose out to the battle for product differentiation.

Quad-Play Services

Quad play is a pre-integrated package of strategies and applications that combines IPTV with voice and video telephony and messaging. The quad-play suite goes beyond triple play by adding wireless/mobile convergence service to the package. Triple play and quad play give subscribers seamless access over multiple devices to personalized services for communication, entertainment, and information. Quad-play customers can watch TV programs and films on demand while downloading music and reading celebrity interviews on a mobile. While a customer could get each digital service from a different supplier, quad-play suppliers offer the consumer all four services as a bundle, and usually at a discounted price.

Features of Quad Play

Quad play refers to mobile/wireless access to services. But mobile/wireless is just another access technology, as are digital subscriber line (DSL), fiber, and cable technologies. It is understandable to think of it as a separate service because the addition of wireless access and mobility applications certainly improves the overall business model (and complexity) of triple-play services. Quad play is the first step in the converged wireless/wireline evolution.

The four services that comprise quad play are video/TV, voice, high-speed data or broadband Internet, and wireless/mobile services.

The quad-play suite goes beyond triple play by adding mobile convergence services. It provides a packaged solution for voice, video/TV, and mobile services over multiple devices (e.g., fixed phone, mobile phone, TV, personal computer [PC]).

Following are the two main features of quad play, which allow users a much better experience than the one offered by triple play:

Remote Interactivity
Combining IPTV and mobility in a quad-play package enables remote interactivity, giving users flexibility in TV and video programming and viewing. Services can be accessed and controlled from multiple devices such as a mobile phone, a videophone, or a PC, allowing users to plan and view programs anytime, anywhere. Services include mobile TV, personal video recorder (PVR) control from a mobile phone, and accessing and displaying media center services on dual-mode phones.

Seamless Transfer over Various Platforms
Seamless transfer enables a voice call on your home phone line to be transferred seamlessly to your mobile as you drive to work or a movie on TV to be paused mid-show and then watched on a wireless personal digital assistant (PDA) as you relax in the garden. Imagine having a cell phone conversation with two or three friends and simultaneously sharing a video of the cricket match you are attending. Then imagine that all of these things can be done with a single account, on a single login, with multiple devices over different types of access networks. These are only a few examples of the quad-play services that can be accessed by users anywhere at anytime.

Reasons for Growth of Quad Play
Increased Value Addition for Customers
Increased value addition for customers has triggered the growth of quad-play communications offered in which the service provider is selling a package of services on a single bill with a set of common features and discounts that encourage the consumer to buy all of those services from a single provider. The major objective behind this bundled service provision is to lock in subscribers by offering them lower prices than consumers would pay for each service separately. This in turn gives a boost to the number of new customers added as well.

Quick and Cost-Effective Service Provision
The quad-play suites provide a quick and cost-effective way for service providers to enter the bundled services market and capture a greater share of subscribers' total communication and entertainment needs. Pre-integration of solutions simplifies deployment and reduces time to market, while synergetic applications enable service providers to create innovative services that generate new revenues and encourage customer loyalty. The triple-play and quad-play suites can be fully integrated with converse billing solutions. Thus the service providers get higher returns and the customers get all the services on a single platform.

Boost in Customer Lifetime Revenue
The quad play dramatically boosts customer lifetime revenue and is 50 percent more valuable than the triple-play suite. This leads to a significant drop in churn rates, implying that a significantly smaller number of subscribers are leaving the service cover provided by one service provider in favor of another bundle. This is because subscribers no longer need to go to different service providers every time they want to start using a new service.

Benefits for Service Providers

Quad play offers communications service providers several promises that are hugely attractive. The opportunity to augment low-margin voice call revenues with new entertainment services, delivered over a common network infrastructure, offers reinvigorated growth to top-line revenues and perhaps higher profit margins. Expansion into the media and entertainment industries offers the chance to leverage a higher fraction of consumers' disposable incomes and with it the opportunity to shed the staid "utility industry" image, which has becalmed the stock price of most telecom operators since the end of the technology bubble five years ago. These promises are real and achievable. However, service providers must revolutionize several aspects of their customer access networks to fulfill these promises.

Figure 1: *Quad Play in Action*

It also enables the service provider to centrally manage and provision subscribers for voice, video, IPTV, and mobile services. New services can be created, added, removed, or modified quickly, easily, and cost-effectively, enabling service providers to gain a strong market position through flexible and innovative service bundling.

The benefit so far in introducing another service in to the triple-play package seems obvious. Quad play makes it possible to define new cross-technology services such as shared mailboxes, increase customer satisfaction, and boost profits.

The quad-play suite enables delivery of enhanced content to the end user and a communication experience based on ease of use and seamless access to personalized, context-aware services: My service, my content, my community, etc.

Factors Affecting the Success of Quad Play

- Anywhere place-shifting
- Mobile connection to the contents of digital video recorders (DVRs) from the road

- Recorded TV content transfer to laptops, smart wireless phones, and PDAs
- Remote access to PC (or home media center) content
- E-mail, voice mail, video mail, IM accessibility from any device
- Anytime time-shifting
- Access content at the most convenient time
- Digital video recording
- Networked personal video recording (NPVR)
- Individual preferences person-shifting
- Priority definition between incoming calls
- Personal directory access from any device: fixed phone, mobile phone, PC, TV
- Phone book, speed dial, buddy lists, contact lists

Market Outlook

- Analysts predict that strong market uptake for triple play/IPTV and fixed-mobile convergence (FMC)—key components of quad-play services.
- The Multimedia Research Group (MRG) forecasts that IPTV subscribers worldwide will

---

End-User Benefits
- Ubiquitous access to a complete range of easy-to-use services
- One subscription for full range of voice, data, video, TV, and mobility services
  - VoIP, video telephony, messaging
  - Live, time shifting, and interactive TV
  - Video on demand
  - Gaming and entertainment services
  - E-commerce
  - My content anywhere, anytime, any way: at home/away from home/on the move

---

Table 2

---

grow from 4.3 million in 2005 to 36.8 million in 2009.

- According to industry analyst Visiongain, FMC will drive fundamental change in both fixed and mobile industries and the market is set to grow to $74 billion by 2009.

Actual Market Trends

- In Denmark, VoIP minutes outweigh landline voice minutes.
- One out of three U.S. homes will use VoIP in 2007 (ChangeWaveResearch, 11/2006).
- 1.1 trillion short message service (SMS) messages were sent, with $ 50 billion in revenue, in 2004 (Informa5/2005).
- In Japan, more e-mails are sent via mobile than via PC (DoCoMo 2005).
- There are already more than 200 million corporate IM users (IDC).
- Many billions of devices will be connected to the Internet by 2010.

In a very short time, triple-play (fixed voice, TV, and Internet access) and quad-play (fixed voice, wireless, TV, and Internet access) offerings have emerged as the consensual vision for the future of the industry.

Avenues of Growth for Service Providers

In today's world, the user is king and demands personalized services. Successful service providers will be those able to offer lifestyle service bundles to groups of users with similar interests and usage patterns. The quad-play suite is the enabler of these new lifestyle services in a multi-play environment.

Role of Cable Companies

Cable companies will also continue to play a role by adding VoIP–based phone service to existing video and data services and forming partnerships to provide wireless service. Telecom companies are currently fighting for the legal rights to franchise these services and lay down the massive fiber networks they require.

Role of Other Telecoms

To recoup their massive investments, it will not be enough for telecom companies to bring in revenues from bundled services, even in light of the associated retention benefits. The opportunity for telecom companies will lie in positioning themselves as both data networks and providers of aggregated content, and targeted IP–based advertising to stationary and mobile devices will be a big part of this.

Role of Bundled Services

Service providers are increasingly looking at bundling as a means of reducing churn and increasing revenue. They hope to build on their success with double play by deploying triple- and quad-play packages and offering increased services to their existing customer base. Bundling has the potential to revive the fortunes of service providers if they develop packages that are in tune with their market segmentations.

But the real bundled services battle will be in developing and marketing the continuity of service and content across different devices (e.g., TV, PC, telephone, wireless). The fate of the "net neutrality" debate, content aggregation, technology development, and incumbent positioning in these service areas will underpin the competitive advantages that telecom and cable companies each will hold.

Potential Pitfalls
However, these service providers also face significant hurdles along the way.

Price Wars
The major hurdle comes in terms of price wars that break out due to the high price discounts targeted at attracting customers. Many experts believe that certainly in many markets, where these conglomerates moving toward quad-play services are all trying to gain an increasing market share from a limited population.

Decreasing ARPUs
A price war situation will obviously decrease average revenue per user (ARPU). A potential way out of downward-spiraling ARPU is to concentrate on the provision of value-added services that only become possible through the bundling of these four access technologies. The general perception is that, as an increasing number of organizations manage to bundle quad-play services under one brand, competing on access price alone becomes a risky strategy.

Conclusion

Datamonitor predicts that approximately 15 million households will purchase IPTV services by the end of 2007, generating global revenues in excess of $7.5 billion. The industry expects to see a complete transformation in the networks and the service portfolios of local telephone companies (incumbent local-exchange carriers [ILECs] or telecom companies). As telecom companies deploy entertainment video services, they will have to decide whether to use a traditional broadcast overlay network of their own, resell a satellite TV service, or implement video over a high-end IP network. While reselling a

satellite TV service may be the most expedient, an IP–based solution is the only solution that can deliver a truly converged video-voice-data experience.

While the potential of triple and quad play to generate long-term profitability has been widely recognized, adoption has been limited by the high initial investment and complex integration of components. The quad-play suite has pre-packaged the components, simplifying service deployment, reducing integration costs, and accelerating time to market. The pre-integrated quad-play suite is designed to lower the entry barrier to the quad-play market, making it easy and cost-effective for service providers to start offering quad-play services today. Quad-play solution enables service providers to increase revenues and customer loyalty by providing innovative bundles of personalized lifestyle services that can capture a greater share of the subscriber's communication and entertainment wallet.

References

[1] "The Quad Play – The First Wave of the Converged Services Evolution," inCode Advisors, white paper, February 2007.
[2] "Residential Triple Play Market Study," The IP Development Network, 3 March 2006, IPNET.dev.
[3] "The Voice Piece of the Triple-Play Puzzle," MetaSwitch.
[4] "Delivering converged quad-play services with IPTV and IMS," Nortel white paper.
[5] Johannes M. Bauer (2005), "Bundling, Differentiation, Alliances, and Mergers: Convergence Strategies in North American Communication Markets," Quello Center Working Paper 05-03, Department of Telecommunication, Information Studies, and Media Quello Center for Telecommunication Management and Law, Michigan State University.
[6] John Mellis, (2006), "Stretching the boundaries – the new access network," Evolved Networks.
[7] Jeff Gordon, Convergys, "Will Quad Play Be a Home Run or a Strikeout for CSPs? The Answer May Lie in Your Customer Service," Pipeline, Volume 3, Issue 10.
[8] Michael Hopkins (2006), "Here Come the Quad Play Cable Companies (and the Competition): Prepare for a Wireless Component," The Bridge, October 27.

Next-Generation Network Management

Semyon Shur

Director, Solution Engineering Department
TTI Telecom

The NGN Challenge

In today's world, the activation and management of new services is a key aspect in day-to-day telecom operations.

Communication services are rapidly changing. Future services will be very different from those consumed today. Though it is hard to predict how technology will advance and what new services the subscriber will demand, the reality is that service complexity and diversity continue to rise.

Technical personnel today are handling more network elements of different types while simultaneously coping with an increasing volume of content applications and data-intensive services per subscriber.

Today's services utilize hybrid networks from the customer-premises equipment, through the operator access network, over its core network, and through to the other end. The services can traverse multiple networks and be separated to many different network elements (NEs).

Today's networks are undergoing an evolution of new services and new network infrastructure. For example, the asynchronous transfer mode (ATM) network, which used to be one of the main transport networks, is being pushed aside and replaced by a multiprotocol label switching (MPLS) core in most vendors. In the voice domain, voice over Internet protocol (VoIP) technologies are on the rise, varying from Centrex technologies in the business and enterprise world to Digital Enhanced Cordless Telecommunications (DECT) services in the residential market.

All of this transformation means that operators require a much more complex management—on one hand, they need to manage new infrastructure and new services, while on the other hand, they still need to manage the legacy infrastructure (which might very well be deployed for several years).

Monitoring of IP Technologies and Services

The information available in the network regarding implemented services appears in different formats (e.g., events, XDRs, counters) Ultimately, this information can provide an end-to-end status of the managed service based on the underlying network.

When looking at the new service offering in today's environment, several main service areas include the following:

- VoIP
- Video streaming
- Virtual private network (VPN) and virtual local-area network (VLAN) services
- Fiber-to-the-x (FTTx)
- Metro Ethernet services

Figure 2 illustrates parts of the new networks the above services utilize.

Another aspect of today's network is that the end user is much more involved in the life cycle of the service they purchase. Customers want the ability to control many aspects of the provisioning cycle (e.g., change requests for bandwidth, additions to the service) and need access to management information for their services (e.g., a predefined set of available reports).

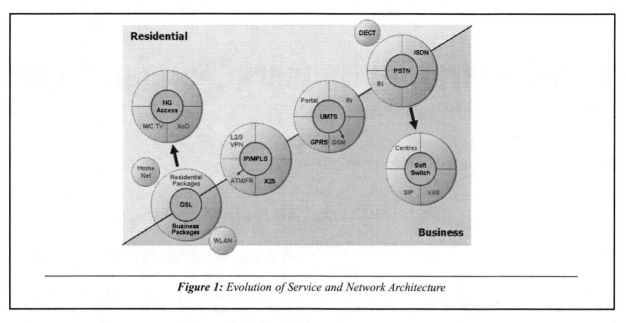

Figure 1: *Evolution of Service and Network Architecture*

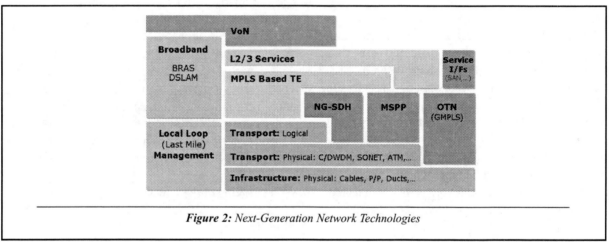

Figure 2: *Next-Generation Network Technologies*

Management of Multiple Services Sharing a Unified Infrastructure

The main challenges in managing these services are related to the fact that a service no longer rides over a single legacy network but traverses multiple networks.

Therefore, the service performance and status is a result of different external data representation (XDR) types and operation measurements obtained from the various networks.

These challenges include the ability to do the following:

- Collect large amounts of data
- Correlate the data and associate it to a specific service and service instance

- Rapidly handle deployments of new services

The NGN Solution – Integrated Management Functionalities for Quadruple Play

The ultimate answer to the above challenges lies in the ability to deploy an operations support system (OSS) solution that will be able to address all of the above challenges while keeping operational expenses (OPEX) and capital expenses (CAPEX) low.

Quadruple play is a growing market requirement of carriers that offers a combination of traditional and wireless telephony, video, and Internet access. Offering several services from one service provider means that end users have a central point for subscription as well as infrastructure. Carriers using

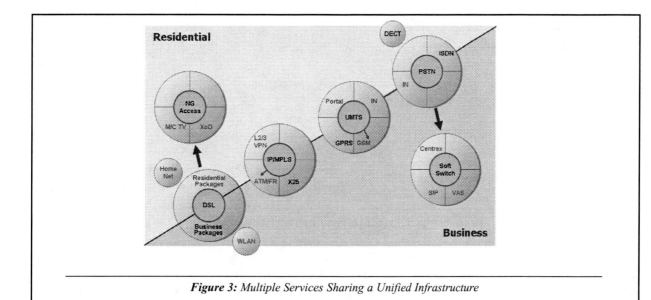

Figure 3: Multiple Services Sharing a Unified Infrastructure

quadruple play sharing unified infrastructure can also decrease management costs and increase their revenues per customer.

Correlated Data Sources

Next-generation services can be managed using the following data source types:

- XDRs—Collection from different sources such as IPDRs, signaling system 7 (SS7) DRs, and LDRs (e.g., IP, SS7, log-sourced detail record).

The collection of XDR allows monitoring of service-oriented performance.

- To get a full picture of quality of service (QoS) for domains such as IP and public switched telephone networks (PSTNs), additional data can be collected optionally using intrusive (active) and non-intrusive (passive) probes. Active probes provide link quality and performance measurements by creating test calls between probes while passive proves provide a single point for quality and performance measurements on all live user traffic calls.

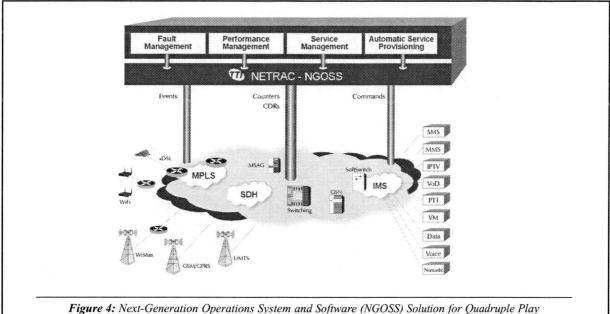

Figure 4: Next-Generation Operations System and Software (NGOSS) Solution for Quadruple Play

- Performance measurements collection—Performance measurements (PMs, or counters) are collected from a variety of data sources such as switches (including ATM switches), digital subscriber line access multiplexers (DSLAMs), routers, gigabit switch routers (GSRs), and transmission equipment.
- Variety of alarm types collection, including the following:
  - Hardware and software alarms from all domains
  - SS7 alarms
  - Performance alarms

In today's evolving environment, end-to-end management and monitoring of services that migrate from legacy networks to NG networks, and quadruple-play services sharing unified infrastructure cannot be based on separate monitoring of the sources listed above. Only by correlating the various sources of service and network information can the end-to-end picture be created.

Conclusion

It has become a serious issue for service providers to face the challenges of rapidly changing networks without compromising the quality of performance management. The growing complexity of managed services and the huge amounts of data requires sophisticated and user-friendly automation tools.

Consolidation among carriers and migration to converged next-generation technologies are occurring at an accelerated pace. Companies need to provide the solutions to fulfill the requirements of managing quadruple-play next-generation services, based on correlation of various data sources and automation of management processes.

Winning in the Age of Convergence: Product Framework for CSPs

Ravikumar Sreedharan
Principal Consultant
Infosys Technologies

Ramakrishnan Ananthanarayanan
Consultant
Infosys Technologies

Ranjit Jagirdar
Senior Consultant
Infosys Technologies

The convergence phenomenon is changing the way communication service providers (CSPs) operate as never before. The falling public switched telephone network (PSTN) revenues, coupled with triple/quadruple play, are forcing the companies to formulate a convergence-led strategy for future growth. The moot point is whether companies have the process and information technology (IT) flexibility to cut across the silos of products/services. The article concentrates on the complex product environment structure across CSPs and proposes a product framework to cater to convergence.

The proposed product operations framework provides a centralized visibility of the enterprise's product propositions across different customer segments and product lines. It also provides a product and bundle creation environment and synchronizes deployment of new products cutting across lines of business. The framework provides a single product master across the organization maintaining both business and product hierarchy. A logical entity for product rules would be the core of the framework. This would be synchronized with the operations support system (OSS) and business support system (BSS) through a service-oriented architecture (SOA). Organizations would need to be cognizant of the impact in bundling too many rules in a single logical entity. The framework bridges the gap between capability and market and also provides support and consistency for product management processes through centralization. It also provides a single point of truth for all the product and offer capabilities.

The new industry phenomenon of convergence across voice, data, and mobility is changing the way CSPs operate as never before. Traditional PSTN providers are feeling the pinch as a result of falling landline revenues. Needless to add, they are often the first ones in the marketplace to formulate a convergence-led strategy for future growth. The moot point is whether companies have the process and IT flexibility to cut across the silos of voice, data, and mobility products/services. The environment is very similar to the silos of assets, liabilities that existed in large retail financial services sectors. This article concentrates on the complex product environment structure across CSPs and proposes a product framework to cater to convergence.

Convergence is redefining the way business works, and consumers are the primary beneficiaries. Convergence provides the ability to mix and match services to cater to the market's needs of flexibility, quality, and value. Convergence is being implemented quickly and has impacted devices, services, and networks. For example, fixed-mobile convergence (FMC) handsets now provide consumers with an ability to swap between their landlines and mobiles. These provide the consumers with the cheapest network available at a given location, and this decreases the cost of communications service for the consumer and pressures revenue for the CSP.

CSPs have identified these external drivers of convergence and coupled them with internal drivers to come up with flexible and innovative models to suit the consumer. Additional drivers include the following:

- Decreased time to market, as CSPs need to move much more quickly in product service rollout.
- A flexible business model, as CSPs have extended their business models to ensure that new services are provided either through networks or partnerships.
- CSPs must increase their focus and intimacy toward the customer in order to reduce churn. One of the must-adopt strategies for service providers to increase customer loyalty and stickiness is the bundling of services (triple play, quad play, or bundling with other partner services such as DirecTV or Dish TV).
- Decreasing PSTN revenue per minute has had a drastic effect on average revenue per user (ARPU). Hence, companies are increasingly providing bundled offerings with value-added services to increase ARPU. In order to support ARPU growth through product bundles and convergent offerings, there has to be increased efficiency in operations in creating and managing the entire product portfolio of a service provider.
- Individual services are being commoditized, leading to decreasing differentiation and, hence, increased competition. The key is in bundling these services into packages.
- Customers are less likely to defect to competitors because of multiple service relationships and the resultant offers such as switching costs are raised in bundle scenarios.
- FMC will act as a defense against pure-play VoIP providers that cannot offer the convenience of a "mobile-fixed" handset.
- New market entrants, including Skype's recently announced fixed and mobile package in

North America, serves to add fuel to the competitive fire.

CSP companies with a PSTN bias are the first ones in the market to start the process of transformation of their existing OSS/BSS systems in addition to laying down next-generation networks (NGNs) and partnering content providers. The complexity of product propositions that are offered in a diverged market (non-convergent) would need to be demystified to cater to the converged market space. The differentiation between individual services would be low as a result of increased competition, and this would mean that companies would need to differentiate through branding, bundling, and complex but flexible services. Challenges in new product introduction have been an issue to these companies from time immemorial, but convergence forces the need to address them now. The good news is that industry body standards, technology trends of Web services, and SOAs also give the companies a scope to look at their legacy-based architecture and take the first steps in their path to facilitate convergence.

Current Model

In a typical BSS environment today, a new product introduction involves heavy dependence on manual and semi-automated processes. As depicted in *Figure 1*, marketing always places heavy demands on the IT organization, and the IT organization struggles to meet the marketing demands on time. Each system in the IT landscape has its own product data and associated functionality.

One of the biggest challenges is the existence of different product IDs for the same product across multiple systems. A mapping of the various product IDs is to be maintained somewhere in the IT landscape.

Sales and marketing—The sales and marketing application administrator updates the system with details around the new product, promotions associated with the product, pricing details, and commission details. This is typically done through a file upload or a manual configuration on this platform.

Order management—The order management application administrator updates the system with the new product and bundle information. This will include generation of new product codes and the setting up of the compatibility rules for the order. Any rules pertaining to eligibility of the product also have to be set up. These are typically done through front-end configuration or through a file upload.

Billing—The billing platform is the backbone or the core of the product catalog. Product catalogs are largely driven by what billing systems support. Billing configurators perform new product ID generation, creation of the price plan, rating configuration, and configuration of journal entries. Configuration in the billing system is also typically achieved by manual entry or by file uploads.

Customer-relationship management (CRM)—The CRM system administrator performs updates on compatibility rules, price lists, and activation/order codes. If the new offer has product bundles, the bundling has to be visible in CRM so that the contact center agent or any other CRM user can position the bundle effectively to new customers or prospects.

Self-service—The self-service administrator creates the products along with associated rules. The bundling and product configuration has to be accurate on the self-service platform, as this is used by the customer directly and there is absolutely no room for back-office manual work-around. Self-service is the most important system from a customer experience perspective, hence the need for accurate product bundling and pricing information.

Provisioning—The provisioning layer administrator creates new order codes and technical products. If the new bundle introduced by marketing involves configuration and setup of new technical products, then the provisioning layer becomes a touch point in the introduction of a new offer or product bundle.

Mediation—The mediation layer needs to maintain the rates and certain billing information. Introduction of a new bundle/product could potentially have a mediation touch point. The rates and billing details need to be configured on the mediation platform.

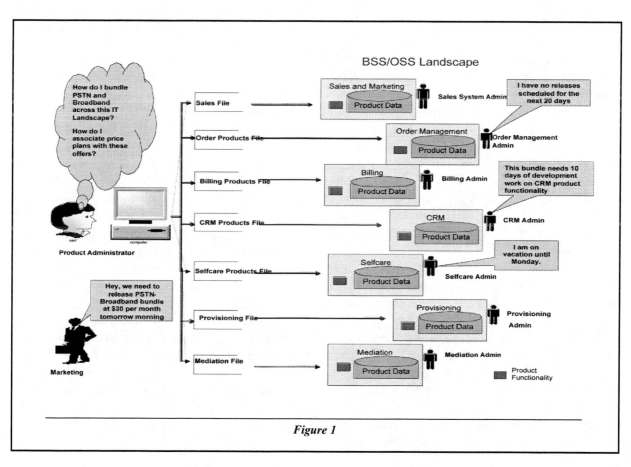

Figure 1

Today's Scenario

A typical product introduction can take anywhere from 80 to 100 days due to the manual overheads and the coordination involved across a large number of stakeholders. This is a huge overhead because it erodes the first-mover competitive advantage and delays revenue generation.

The problem is expected to be further compounded in the future due to innovative and dynamic product bundling needs imposed by convergence. For example, today an introduction or change of a price plan from concept to market takes an average of 80 to 100 days. If the example was that of the introduction of a convergent offering such as a PSTN and mobile line (FMC) with differentiated prices based on the location of the subscriber during a call, coupled with a Christmas offer of a 50 percent discount on the first month's bill, but only for purchases in the London district of the United Kingdom, the implementation in the BSS would be a nightmare! By the time a program of this nature is implemented, significant revenue is lost as a result of the slow time to market.

Billions of dollars in revenue and profits go unrealized because service providers cannot quickly respond to market opportunities and competitive dynamics. Research shows that implementing product life-cycle management methodologies in other industries can achieve a 40 percent reduction in product development time. This efficiency represents significant cost savings and has a positive impact on revenue goals (source: TeleManagement Forum TMFC1836, The PLM Problem).

How Do Communication Service Providers Address Today's Challenges?

To address the challenges with today's model, CSPs need to perform routine "architectural health checks" on the overall health of their BSS/OSS architecture with particular focus on the systems that house the product catalog. They need to constantly evaluate whether the changes and the growth of their BSS/OSS architectural landscape are in alignment with future business needs and growth plans around new product bundles and offers. This assessment is ideally performed by an external consultant/vendor body that brings in specialized expertise in the area of BSS/OSS systems integration and architecture study. A periodic health check in this regard would be useful, considering the dynamic nature of the market and the business needs.

The consultant/vendor body would typically perform a study of the as-is architectural landscape and as-is business processes around product management and selling. They would also assess changes to selling specific business processes and SLAs that would take place in the immediate future as a result of new bundles and product offers. At the end of the study, the consultant/vendor would recommend any changes required to the BSS/OSS landscape around products management from a strategic perspective to align with business goals.

The CSP should be careful to bring in independent and neutral vendor consultants to perform this study, i.e., consultants who would not have any specific bias or interest toward the promotion of a particular application or product. The consultants should be able to recommend optimal solutions that meet the specific environment and business needs of the service provider.

A typical framework for managing products in a converged market place is discussed below.

Product Framework for a Converged Marketplace

Convergence is a catalyst for organizations to review their product portfolio strategy. By deploying next-generation capabilities and establishing new content and services partnerships, CSPs are moving away from selling simple products or service capabilities and moving toward complex and bundled products.

For CSPs with multiple services across voice, data, mobility services, and content services, the only way to achieve the benefits of consolidation (product bundling for cross-sell/upsell) is to ensure that the product application landscape is simplified. This would not only shorten the time to market and cater to convergence, but also reduce costs in manual/multiple processes around products.

Characteristics

The product framework provides a centralized visibility of the enterprise's product propositions across customer segments and product lines. It also provides a product and bundle creation environment, synchronizes the deployment of new products cutting across lines of business, and provides a flexible workflow to coordinate the teams involved. This workflow helps in automating the following tasks to formulate the rollout of a new product/price plan:

- Creation, review, authorization, and approval of the product structure
- Modeling and testing of the product structure
- Versioning of the product structure to cater to any changes
- Rollout after final authorization for use in production

The business process flow would involve various levels of the organization across user communities. This flow can be managed effectively through a workflow in order to minimize process deviations.

The framework provides a single product master across the organization, maintaining both business and product hierarchy. A logical entity for product rules would be the core of the framework. This would be synchronized with the OSS and BSS through an SOA. Organizations would need to be cognizant of the impact of bundling too many rules in a single logical entity. The framework bridges the gap between capability to market and provides support and consistency for product management processes through centralization. It also provides a single point of truth for all product and offer capabilities.

The framework encompasses interactions between the logical product repository and OSSs/BSSs. In the telecom environment, there are many interactions between systems in the OSS/BSS space. The system(s) that make up the product framework would typically interact with the following OSSs/BSSs:

- CRM system—Interaction with the CRM system would typically be needed to support the

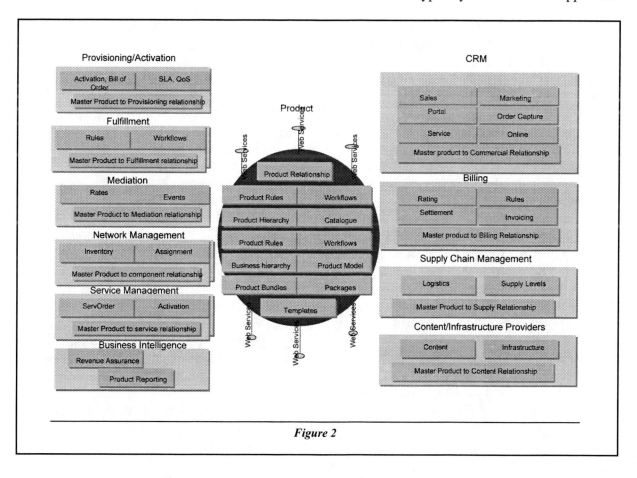

Figure 2

sales, marketing, order capture, service management, and on-line capabilities that are supported by the CRM system in a CSP.

- Billing system—Interaction with the billing system would be required to enable the billing system to support rating, payment and compatibility rules, settlement, and invoicing.

- Supply chain—Maintaining a product-to-supply relationship ensures that available supply levels are adequately tracked, leading to minimal gaps in the demand-to-supply chain. This would require updates on new product information on the strategic supply chain platform from the products framework.

- Content/infrastructure providers—New offerings, especially in the 3G space, include content such as mobile TV and entertainment services for which CSPs join up with third-party service providers. The product framework can facilitate integration into the systems of the third-party service providers in order to achieve real-time synchronization of new content-related products. An interaction of a similar nature can be established with infrastructure and network providers.

- Provisioning/activation system—The introduction of new products through the product framework could entail setting up the technical product(s) in the provisioning/activation system. This would entail interaction between the product framework and the provisioning/activation system.

- Fulfillment—Any new product or product-related rule toward fulfillment may need to be set up in the fulfillment system through an interaction between the product framework and the fulfillment system.

- Mediation—The mediation system plays a key role in mapping the events and rates associated with billing products. Any new product introduction or existing product modification could need interaction between the product framework and the mediation system for necessary data setup.

- Network management—The relationship between the product and the network component provides a crucial link for managing the network. Any new product that involves a new network product or update of network parameters for an existing product would be facilitated by an interaction between the product framework and the network management layer.

- Service management—Service management would involve providing after-sales support and service on products. Any new product introduction would involve synchronization with the service management platform to facilitate real-time or semi-real-time product and service updates.

- Business intelligence—A centralized product framework can facilitate support for product data mining for marketing research and forecasting purposes. This can be achieved by mining data in the correct format and feeding into the business intelligence platforms.

Key Benefits

The following are some key benefits the CSP could accrue as a result of building and maintaining a product framework in a converged marketplace:

- Improved customer experience due to better serviceability and capabilities, which provides value for customers.

- Consistency and accuracy in the product management processes across multiple business units and OSS/BSS landscape.

- Significantly reduced operational costs due to streamlining the maintenance of multiple product functionality in different application landscapes. This also decreases the IT costs of taking a new product to market.

- Decreased revenue leakage by improving business and systems processes relating to product and price changes.

- Improved time to market of services/products and marketing assistance in innovatively increasing the wallet share of consumers.

- A consistent product description across the organization with the ability to distribute real-time product information across the OSS/BSS platform.

- Grouping of products and offers with complimentary products and bundling of complex products that can facilitate ease of after-sale service.

- A single point of product truth and hence a master for the rules, workflows, product hierarchies, and catalogs.

• A standardized product model across the entire organization and increased accuracy of data in the application landscape.

If CSPs would like to survive and thrive in the convergence wave, they must streamline their product management processes. This is the best way for the organizations to cut costs, fend off competitors, retain customers, increase wallet share, and minimize their losses from declining fixed-line voice revenues.

In today's scenario the centralized product catalog will go a long way in helping service providers increase product management efficiencies. However, with advancement in SOA and product organizations increasingly moving toward Web services–based technologies over the next five to 10 years, even the need for a centralized product catalog would need to be re-assessed after this period as part of the routine "architectural health checks" prescribed earlier.

Quad Play: A New Telecom Service Trend

Hung Tuan Tran

Senior Technical Consultant, Global Business Development Sector
FPT Information System (FIS)

Introduction to Quad-Play Service

Technology developments in recent years have opened the convergence paradigm, a promising trend in telecommunications networking. In the rough sense, the term "convergence" refers to feature integration and consolidation. For example, to integrate and consolidate the features of various endpoint devices into one single device means device convergence; to enable a common technology architecture baseline for cooperation of a variety of wired and wireless, fixed and mobile networks means technology convergence; and to consolidate various services into one single service delivered to subscribers represents service convergence. In particular, service convergence now enables telecom providers to provide to subscribers not only traditional voice (telephony), but also a rich set of multimedia applications packaged in one service offering. Quad-play service is a representative of such convergence, providing in a single service package the features of voice, video, and data delivery as well as mobility.

Quad-play service in fact is the next development step from triple-play service. The latter one itself is already a converged service, leveraging the existing IP infrastructure and IP technology to ensure voice, video, and data communications service to subscribers. Triple-play service has been fiercely dealt with and released as a real-life offering from the years 2004–05. It should be noted that the fact that IP technology and architecture are utilized to transmit voice, video, and data as in triple-play service is not considered new anymore in today's telecommunications field. In essence, the distinguishing and innovative point of triple-play service is that all the components (e.g., voice, video, data) are bundled into a single service package. This single package thus provides the subscribers with a rich set of multimedia applications having very attractive features that can be personalized to tailor to the best convenience of the subscribers.

In the last years, new developments and achievements in the access technology areas bring even more opportunities to better satisfy demands of subscribers. In particular, triple-play service now can be expanded with the mobility feature to form the so-called quad-play service. In the paradigm of quad-play service, subscribers can use their wireless devices (e.g., laptop, personal digital assistant [PDA], mobile phone) to access and use triple-play service features with the same quality experience as if they were sitting at home using wired end devices (e.g., telephone handset, television).

Triple-Play Service: A Descriptive Look

To better shed light on the characteristics of triple-play service and on the features it offers from the perspective of subscribers, let us go to a somewhat deeper description of Internet protocol television (IPTV) service. Being the most representative example of triple-play service, IPTV has been already launched recently by a number of telecom service providers in several regions in the world. Using the same Internet connection (which also accommodates traditional telephone and data delivery service) to end users, IPTV brings to them value-added television and video services.

In general, to use IPTV, the subscriber first of all needs a special device called a set-top box (STB). This STB connects to the Internet via cable or ADSL connection and to the traditional TV set. With the STB and the normal television set, the subscriber can use IPTV service. *Figure 1* shows the basic set at the subscriber site for IPTV service.

Using IPTV, a subscriber can watch TV channels as he or she would by using cable TV service. They can also use video-on-demand service, selecting the film from a given list. Film selection can be done in an interactive manner on the TV screen using the remote control. Also interestingly, IPTV service may have screen-on-screen display, meaning that a number of smaller TV screens are placed side by side on the TV screen. Each small TV screen corresponds to one channel so that it facilitates the subscriber in choosing which channel to watch. Beyond that, IPTV provides the rich set of integrated voice and data applications. Interactive chatting between IPTV subscribers is possible on the TV screen, with a full set of features (e.g., sharing emotions, avatar usage) similarly to any popular chatting program such as Yahoo Messenger or Skype. Chatting can be initiated while watching TV, allowing subscribers to share their opinions and feelings about the TV program/film being watched. Features for telephony applications are also included in the IPTV service. For example, an incoming call alert will appear on the TV screen (accompanied with or without sound, accordingly to subscriber setting). If the subscriber would like to answer the call, he or she can use the traditional phone set or the speaker phone. The subscriber can also reject the call or send it to voice mail. *Figure 2* shows some typical IPTV screens.

As already mentioned, IPTV service has started to be launched across the world, especially in developed countries such as the United States. For example, in Switzerland, the telecom provider SwissCom began to offer IPTV BlueWin service on November 1, 2006. The BlueWin service delivers 100 TV channels, 70 radio stations, 500 films on demand, sports programs, and around 30 Teleclub channels.

Quad-Play Service from the Perspective of Service Providers

For service providers, the quad-play paradigm does not simply mean offering a new bundled service.

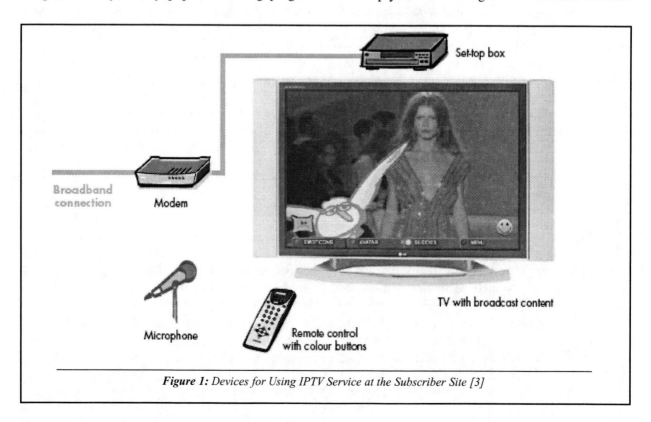

Figure 1: Devices for Using IPTV Service at the Subscriber Site [3]

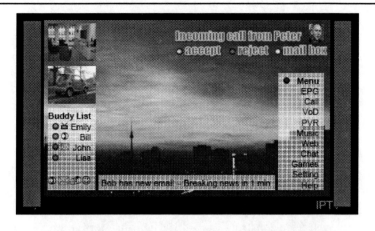

Figure 2: Some Typical Screens of IPTV Service

Under the contiguous pressure of revenue and business benefit, the providers must achieve a thorough analysis on market demand with an outcome that will lead to the right decisions on technology and solution investments for quad-play service. The most interested players in this quad-play field are the telco providers and the cable operators, since they can exploit their existing infrastructure (e.g., telephone network, IP network, cable network) to achieve an IP–based converged service such as quad play. Before deciding to really go for quad-play service provision, it is very important to perform operational tests to identify and fix potential problems that could cause subscriber dissatisfaction. This is even true for the triple-play service road map; for example, Swisscom went through this process in [4] before launching its IPTV service. In reality, end users are only motivated to subscribe to

IPTV if they contemplate that the quality of service (QoS) is worth the money they pay. Moreover, according to market analysis and surveys, it is commonly known that end users expect that the fee they pay for IPTV is less than the total fee of three separated services of phone, video, and TV, and the expected discount is about 10–20 percent [5].

At the time of this writing (2007), the subscription for triple-play service is not much cheaper than the total fee for the three separate services of voice, video, and data. This is also anticipated in case of quad-play service, at least for the first few years after its launch. Consequently, in a fairly competitive market, the overarching factor for the success of subscriber acquisition and market expansion for quad play is the service quality that the provider can deliver to the end users. This is why providers should spend time, effort, and money on a careful testing process before going live with quad-play service.

Figure 3 depicts a general architectural view of x-play service (which is true for triple-play and quad-play service), including the associated main technologies and devices [1]. As can be seen, IP technology and related technical points (e.g., infrastructure, transport technologies, voice and video traffic management, billing methods), cooperation between IP, telephony, and content networks are the main building blocks of x-play service construction.

Basic Technology Components in Quad-Play Service

To form the complete picture on the quad-play service on the IP platform, let us briefly review its key technology components and the inherent technology issues.

IP Platform

The booming spread of the Internet and its underlying IP–related technologies in the early 1990s contributes to the fact that IP platform today has been broadly deployed across the world. Transmission control protocol (TCP)/IP has become a de facto protocol, ubiquitously adopted as a transmission platform for a variety of data types. The telecommunication industry in fact is trying to utilize and exploit the existing IP platform to accommodate a number of value-added services to the end users.

VoIP

One of the four key components of quad-play service is the transport of voice traffic in the IP platform, often referred to as voice over IP (VoIP). At first hearing, this might seem to be a simple task. We only need to convert voice signals to digital form, encapsulate the bits into IP packets, and push the packets to IP network for delivery to the end users. However, in reality this is not so simple, regarding the deployment and implementation of VoIP. On one hand, the IP network operates in the best-effort mode without any guaranteeing quality for the voice packet delivery. On the other hand, VoIP traffic usually requires strict QoS. The phone conversations must reserve the eligibility, i.e., technical issues cannot in any way be the reason that lead to the situation where participating partners

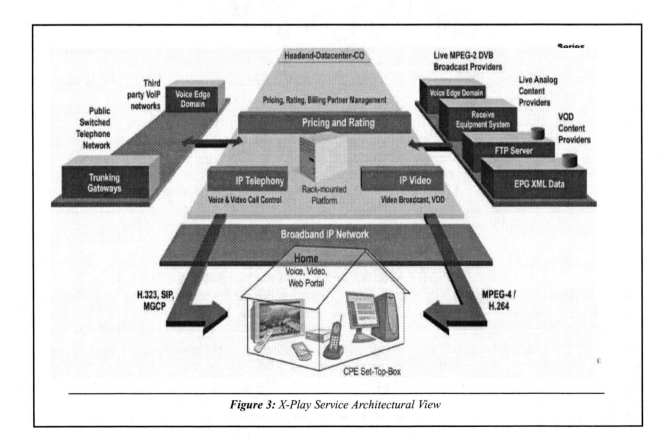

Figure 3: X-Play Service Architectural View

cannot understand each other during the phone call. For other voice applications such as music-accompanied voice in video, the quality requirements are even more rigid. The quality of music and accompanied voice must not suffer distortions or losses.

From the technical point of view, quality assurance for VoIP applications (be it phone conversation, music, or accompanied voice in video) means that QoS mechanisms must be worked out and deployed onto the IP platform. These mechanisms will help minimize the packet loss rate, packet delay, and jitter encountered by the voice traffic, thus keeping the quality of VoIP applications at the desirable level. A number of such mechanisms have been a focus for research and development and even been integrated into market-ready telco products. Exemplary mechanisms and technology areas for VoIP QoS include connection admission control schemes, traffic differentiation and prioritization policies, traffic engineering, bandwidth management, optimal multicasting communications, optimal media server location and utilization, and application-specific optimizations.

Video over IP

Video over IP is the second key component of quadplay service. Similar to VoIP applications, video over IP also has strict requirements on quality assurance. Movies and accompanied voices must have quality that is at least commensurate or even better than that of the traditional television channel (received via antenna or cable operator) quality. In addition, in case of live broadcast programs, realtime transmission and synchronization are stringently required. The scenes seen by the end users must be synchronized in time with the real event that is being broadcast live, and almost no delay should be tolerated.

Video traffic, however, is different from VoIP traffic in that a video traffic flow would consume a bandwidth in a range of several Mbps, while a VoIP flow only takes some tens of kbps. This difference implies that special technical solutions must be prepared in addition to the QoS mechanisms that work well for VoIP. In other words, it is true that QoS mechanisms/technologies developed for VoIP traffic (mentioned earlier) are to some extent still applicable to video traffic. Nevertheless, due to special characteristics of video traffic, other mechanisms are needed as well. For example, issues related to video encoding (e.g., Moving Pictures Experts Group 3 [MPEG–3], MPEG–4), content networking mechanisms, and load balancing mechanisms are some main topics concerning video traffic transmission over IP.

Data over IP

Transmitting data over the IP platform has become a familiar task in today's technology. In fact, all the activities of Internet users rely on data transmissions taking place in wired local-area networks (LANs), wide-area networks (WANs), or wireless networks. The primary issue of data transmission on the IP platform is that of enhancing quality of the transmission service and tailoring it to the actual requirements. In this field, various mechanisms have been developed or are being developed, and some have already been integrated into telco products widely available in the market. Key examples for such mechanisms encompass congestion control, congestion avoidance mechanism, traffic engineering mechanisms, protocol enhancement methods, and many others. In particular, one of the hottest issues today seems to be the security issue, dealing with the methodologies, policies, and mechanisms to assure confidentiality, integrity, and availability (CIA) for data transmission in the IP platform.

Mobility and Wireless Access

The customer demands trigger the need for mobility of multimedia service, i.e., to make the service available anytime, from anywhere. Today, this trend is gradually materialized and lately is more intensively promoted with the involvement of two technology breakthroughs: worldwide interoperability for microwave access (WiMAX) and IP multimedia subsystem (IMS) platform.

IMS is a common networking platform for multimedia service delivery to end users. It is a collection of technical specifications and standardizations developed and controlled by the third-generation partnership project (3GPP). IMS relies on two key components: IP technology and the signaling protocol session initiation protocol (SIP). As mentioned earlier, IP nowadays is a de facto technology in telecommunications. Besides, SIP is a signaling protocol gaining more and more deployment and acceptance, especially in core networks. By relying on IP and SIP, IMS serves as a single common platform, enabling the cooperation and interaction between various networks with different base technologies: wireless LAN, multiprotocol label switching (MPLS) network, Global System for Mobile Communications (GSM) network, Universal Mobile Telecommunications System (UMTS) network, and so on. Thanks to this cooperation enablement between networking platforms achievable with IMS intervention, end users could use multimedia service anytime from anywhere, independent of their actual location, their actual access technology in use, and their end-device type. Putting in other words, IMS indeed brings in the opportunity to achieve service convergence, network convergence, and device convergence, as illustrated in *Figure 4* [1].

With the introduction of IMS platform, adding mobility to the feature set of triple-play service to form the quad-play service becomes possible. It means that if the subscriber owns an end device (e.g., laptop, PDA, mobile phone) supporting at least one access technology (e.g., WLAN, digital subscriber line [DSL], Ethernet, general packet radio service [GPRS], UMTS, wideband code division multiple access [WCDMA], WiMAX) he or she can have access to and can utilize voice, video, and data service included in the bundled service at

anytime, from anywhere. In particular, the fast maturity process of WiMAX technology will speed up the presence of quad-play service in the market, making the deployability of quad-play service technically solvable.

Although WiMAX and IMS trends are more than promising from the theoretical point of view, in reality (as of 2007), the real deployment and their coverage is still limited. Real-life deployment of products and architectures are exceeded by research and development and testing phases. Mostly, processes on testing and piloting the operation of IMS and WiMAX are in progress. Issues that draw extensive attentions from both research and industry communities include network performance, interoperability of networks in the IMS platform, QoS, and security issues.

Rollout of Quad-Play Service in Reality

Market analysis and experience show that the demand for triple-play service, in particularly for IPTV, is increasing. The worldwide survey from Canalys company reports that the number of IPTV subscribers in 2006 was already 3.6 million (www.iptv-news.com/content/view/1018/64). However, the distribution is IPTV subscribers is not even, mostly concentrated in the developed countries. The above number of subscribers for IPTV is predicted to reach 40.9 million by 2011, according

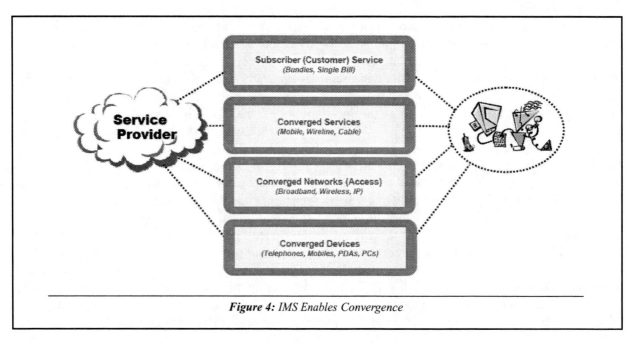

Figure 4: IMS Enables Convergence

to the market analysis company Strategy Analytics (www.iptv-news.com/content/view/1012/64). These market analyses indicate that triple play itself is already a promising service from the business point of view. Thus, once it is has been expanded with a wireless/mobility feature to become quad play, the market potential is expected to be even stronger. On the other hand, in the less developed countries, today's situation is that the service market is not yet fully exploited, even for triple-play service. For example, in Vietnam, among three providers having permission for delivering IPTV service, only the FPT Corporation has launched it, with several service packages covering TV channels offering video on demand, chatting, on-line news, and on-line music. The other two providers, VNPT and Viettel, have not started the IPTV service yet. There is therefore a lot of room for market expansion in such countries, for triple- and quad-play services.

In the world, a lot of quad-play development road maps have been announced and are currently in progress. Some typical examples (until the beginning of 2006) of quad-play service providers are [2]: SBC and Bell South (a telephony service provider in the United States); France Telecom (which currently has 200,000 subscribers for broadband television service and is expected to have 1 million subscribers in 2008); and the alliance of big companies in the United States, including Sprint Nextel (the third biggest mobile provider in the United States), Comcast, Cox Communications, Time-Warner, and Advanced NewHouse (they are cable operators), which have invested $200 million for quad-play deployment.

In the next coming years, we will certainly witness further developments and move forward with quad play, a new trend in telecom service.

References

[1] M. Katz, Creating Quad and Triple Play Solutions for Operators, October 2006.

[2] The Quad-Play – the First Wave of the Converged Services Evolution, inCode's white paper, February 2006.

[3] T. Coppens, F. Vanparijs, K. Handekyn, AmigoTV: A Social TV Experience through Triple Play Convergence, Alcatel technology white paper.

[4] The Battle for Broadband, IEEE Spectrum, January, 2005, pp 25–29.

[5] D. Schmidt, D. Kamarga, Economic Drivers for IMS–based Converged Service, Siemens white paper, November 2006.

Building the End-to-End Media Platform

George Tupy

IPTV Marketing Manager
Cisco Systems

A Holistic Approach to Delivering Quadruple-Play Services

Charles Dickens once wrote, "Change begets change. Nothing propagates so fast ... and what was rock before becomes but sand and dust." Although Dickens could never have imagined what the world would look like 150 years after he wrote those words, the sentiment must ring especially true to the present-day service providers who are seeing their industry transform virtually before their eyes.

Today's providers of voice and data services, wireless services, and television built hugely successful business models around delivering one or two services extremely well. However, service providers now recognize that the days of delivering voice, video, data, and wireless services as distinct offerings, with each delivered over its own network, accessed using its own device, and billed as a single subscription, are rapidly fading away.

To succeed in the future, service providers will need to deliver all types of rich media to all subscribers. Even more important, they must develop the capacity to deliver media to a wide range of fixed and mobile devices and be able to provide a consistent, high-quality experience across all environments. Ultimately, they must transform from traditional providers of access-based services to all-inclusive "experience providers" that can offer the full quadruple play of voice, video, data, and mobile media to subscribers anywhere, anytime, on any device.

Although traditional wireline service providers have well-developed strategies for delivering voice and Internet/data services successfully, video represents an entirely new challenge and more complex proposition. However, it is ultimately the video experience that will promote and differentiate the offering in tomorrow's quadruple-play marketplace. Although today's cable operators currently have a head start in video services, the reality is that the changing media landscape will present new challenges to them as well, as subscribers demand new types of rich media experiences that go beyond the capabilities of today's cable video infrastructures.

Cisco envisions a future in which subscribers use service providers' media networks in the same way they use the Internet today—as a fully interactive environment. In this paradigm, passive (and even basic on-demand) viewing of video entertainment will be supplanted by myriad next-generation applications that the service provider will have to support. Indeed, in tomorrow's media landscape, video will be the prevalent type of content used in a broad range of interactive, user-generated, and community-oriented applications that subscribers will expect their service providers to support.

What do service providers need to understand in order to create a differentiated media experience and build a robust, highly scalable foundation for delivering the applications of tomorrow?

From a top-level, end-to-end system standpoint, the answer can be broken down into the following distinct but complementary domains:

- Service providers need tools to effectively define the media experience in the headend and content-on-demand system (CDS).
- Then, the service providers have to preserve that experience as it is delivered to the subscriber over the intelligent Internet protocol (IP) network.
- Finally, the service providers have to realize that experience exactly as intended in the customer home.

Each of these domain requirements presents its own unique challenges. However, at Cisco, we believe the way to address all of them lies in building an intelligent, media-aware service network that can integrate many types of content and services into a single, comprehensive, end-to-end system based on the intelligent IP network infrastructure (*Figure 1*).

Defining the Media Experience

The media experience initiates and ultimately is defined in the service provider's headend and CDS.

The video headend is the defining factor for the visual aspects of media experience that is complemented and enhanced with the "interactivity" aspect of CDS. In the define domain, service providers need the following proven technologies and strategies for media content:

- Acquisition
- Processing
- Encoding
- Content Delivery
- Management

Service providers have to use broad expertise to help ensure that analog, digital, and IP technologies closely interoperate to produce the desired quality of experience (QoE) end result.

Achieving this is no small task, especially for wireline carriers embarking on video services for the first time. Video headends encompass a wide range of technologies and formats and demand a wide range of skill sets spanning the whole spectrum

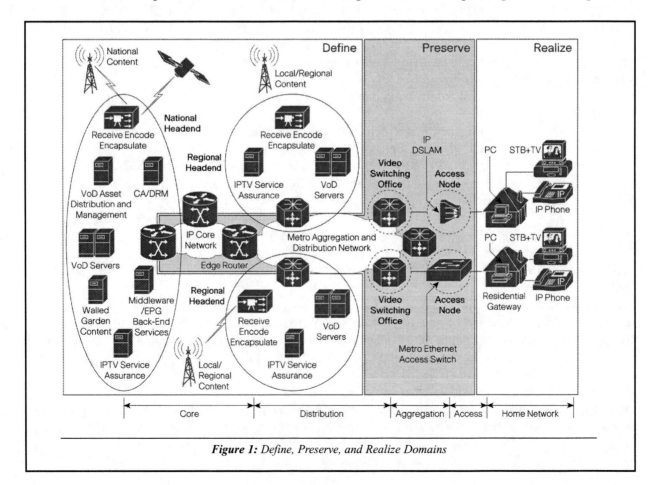

Figure 1: Define, Preserve, and Realize Domains

from the latest Moving Pictures Experts Group (MPEG) encoding to radio frequency (RF) technologies.

Acquisition

Video acquisition alone can encompass satellite, off-air, and fiber sources, each using its own format and encryption protocols. The acquired heterogeneous media content must be decrypted, converted, and multiplexed into a single video service, which is a complex challenge that requires many components to work together. Since every service provider offers its own menu of channels and services, with each requiring a unique mix of technologies, the video acquisition component of the headend must be built from the beginning as a customized solution. For the most part, the headend must be designed, tested, and assembled as a customized solution for each service provider, depending on the specific mix of content and services that will be delivered.

The acquisition segment must be designed with the appropriate redundancy and backup/failover capabilities needed to satisfy the service provider's uptime goals, which can be quite stringent considering the centralized nature of the quad-play video headends.

Processing

Processing media for distribution has become a particularly complex challenge. In the past, when all subscribers accessed similar types of media on the same type of device, the process was relatively straightforward. Today, the situation has changed significantly, and in emerging quad-play environments, subscribers might be just as likely to access content from a three-inch wireless phone display as from a 100-inch high-definition television. Service providers need state-of-the-art capabilities for transrating and transcoding capabilities to optimize content for any format or screen size and provide a consistent high-quality experience regardless of the way media is accessed. To create a more profitable offering, most service providers also want to build capabilities for targeted, even personalized, ad insertion, which might require real-time muxing and remuxing of thousands of simultaneous streams from a variety of sources, further increasing the complexity of media processing.

Encoding

Using the right tools to encode media—especially video—is the most important part of the "define" domain. More than any other factors, it is the video-encoding technology, especially the implementation of that technology, that determines the quality of the media experience. As demand for high-bandwidth content, especially high-definition video, continues to increase rapidly, many service providers are turning to MPEG–4 advanced video compression (AVC) encoding, which requires approximately half the bandwidth of the MPEG–2 technology long used in the existing networks. However, MPEG–4 AVC encoding is an extremely complex process encompassing a much larger set of variables than MPEG–2 algorithms. In addition, MPEG–4 high definition is extremely sensitive to high motion, noise, and rapid luminance changes. Two providers might use the MPEG–4 AVC format to encode the same media source, but depending on the brand of the encoder, the end results subscribers see might be sharply different. Service providers need to make sure they are using MPEG–4 AVC encoding solutions that encompass the full suite of MPEG–4 tools (not just a subset) to optimize encoding for the widest range of visual features and deliver brilliant picture quality at the lowest possible bit rate.

Content Delivery

As subscriber expectations for accessing and interacting with media evolve, service providers are turning to new media distribution platforms that deliver more personalization, localization, and on-demand access. However, modern media services such as time-shifted television, personalized ad insertion, and network-based digital video recorders (DVRs) place huge demands on media networks. Service providers need distribution systems that can ingest, store, manage, personalize, and stream vast amounts of content from anywhere on the network and help ensure that all content is available instantly to any subscriber, anywhere.

Many service providers currently have video-on-demand (VoD) distribution systems in place that actually function effectively as large, centralized video servers. These systems can provide a good solution for basic VoD applications, but they have inherent limitations that prevent them from being ideal distribution platforms for tomorrow's quad-play services. Most notably, they are difficult to scale as the subscriber base and content library grow.

The ideal media distribution system should operate more like a network than a large monolithic server. It should provide three primary capabilities—resiliency, scalability, and load balancing—inherent to a system with multiple distributed resources.

It should also support the full range of emerging media content and applications, not just deliver video to televisions. The system should dynamically bring up and take down new content storage and streaming resources to enable content distribution according to real-time calculations of each title's popularity. Finally, since service providers might need to address many markets and cannot predict how demand will grow, the system should provide the flexibility to serve very large communities, very small communities, and everything in between.

Any successful media distribution strategy also must include intelligent middleware and digital rights management (DRM) solutions to allow subscribers to easily access and legally share media. With middleware and DRM technologies constantly evolving, service providers do not want to get locked into a single strategy. Instead, they should take an open, standards-based, and flexible approach that allows for continuous innovation.

Management

Because of the broad mix of technologies and applications in the headend, service providers will also want to make sure they have management tools that allow them to monitor and control all heterogeneous components from a single system and, ultimately, from a single screen. Since many video hubs do not have on-site staff, carriers must be able to manage all solutions across the service network—including third-party devices—remotely.

Building and operating a video headend effectively require significant rich-media expertise that delivers full, automatic redundancy with autonomous management of backup and failover routines with routing control of each video stream. Finally, a significant aspect of video man-

agement is performance data collection and recording from all network elements as well as performance and trending reports about network availability and performance.

Preserving the Media Experience

The quality of the media experience is defined in the video headend and CDS, but service providers need a quadruple-play network capable of preserving and sustaining that experience and delivering it to subscribers exactly as intended. To do that, the network needs to possess "application-aware" intelligence to distinguish content types and handle each accordingly. The network also must be simple to operate, scale to millions of subscribers, and provide a range of video-specific capabilities, including rapid channel change times, end-to-end video quality experience (VQE), and robust security.

For many service providers, this means adopting an IP–based media approach that can apply IP intelligence to identify and separate types of traffic. IP offers other advantages such as the ability to preserve bandwidth across the network by employing IP quality of service (QoS) techniques to deliver media only to those subscribers who specifically request it. By integrating IP–based content-aware and subscriber-aware intelligence into the network, service providers can also have better controlling, monitoring, and billing of services.

Subscribers might have their own subjective standards for judging the quality of the content they view, but the standards by which media delivery networks are measured are quite specific. For video, the most rigorous media application, the industry standard allows for just one artifact per two-hour movie, which translates to an overall network packet loss of just 10^{-6}. Most video artifacts are caused by brief network outages, so service providers will want a network that uses state-of-the-art resiliency and convergence strategies to achieve sub-second recovery times regardless of where on the network the outage originates.

Because on-demand media distribution systems make it difficult to predict the amount of traffic on the network at a given time, service providers will also want integrated admission control capabilities to police the network and protect service quality even when the system is oversubscribed.

Finally and most important, the delivery network should be a true quadruple-play infrastructure—not just a consumer video network. Most service providers recognize that the most competitive service offerings of tomorrow will bundle voice, data, video, and wireless services and provide an integrated media experience that draws on all of them. Unless service providers want to operate separate networks (and maintain separate management and billing infrastructures) for voice, video, wireless, and business and consumer broadband services, they need a network that converges all types of media over a single infrastructure and can deliver any service to any market over any device.

Realizing the Media Experience

The final component in delivering a compelling media experience is the most important because here the content directly interacts with the subscriber and the experience is re-created though the ears and eyes of the consumer, and a set-top box or mobile video device must deliver an undistorted high-fidelity rich-media experience. If subscribers' in-home and mobile devices do not provide the quality and features needed to fully realize the experience, the "define" and "preserve" segments of the quad-play system will be irrelevant.

Service providers must be able to offer consumers high-performance, easy-to-use options and an intuitive GUI for accessing and consuming media content, both in the home and on the go. In-home set-top boxes are the subscriber's gateway not just for video, but also for the delivery of rich, interactive media experiences. Set-top boxes must deliver outstanding picture and sound quality, as well as support for high-definition content, DVR capabilities, and media center capabilities that allow the sharing of content throughout the home and across multiple devices. Set-top boxes also must be easily upgradable to allow service providers to make continuous enhancements to the media experience over several years and must support robust remote management capabilities.

Building a Comprehensive Media Platform

Clearly, the processes of defining, preserving, and realizing the media experience will draw on very different strategies, each requiring unique technologies and approaches to deliver optimal results. The media networks of the future will require a holistic approach to the end-to-end system and will need to provide many types of content and services to all subscribers. With these types of requirements, using a loose-fitting assortment of technologies from multiple providers is naturally going to produce a media platform that is less efficient and more costly to own and operate.

The ideal quadruple-play platform should be a true end-to-end system that provides built-in intelligence, flexibility, and scalability to support a large and continuously expanding suite of media applications. With the ability to integrate and manage all types of content and services with a single converged infrastructure, service providers have more flexibility to extend the capabilities and scale of that network over time.

Even more important, a quad-play end-to-end system will require much tighter linkages between the media content and the intelligent IP next-generation network (NGN) delivery infrastructure. Ultimately, I believe a tightly integrated end-to-end media platform using a media-aware IP NGN infrastructure offers greater efficiency and manageability across the define, preserve, and realize segments of the infrastructure and provides a more holistic, long-term approach to delivering quadruple-play services.

Service Delivery in a Quad-Play Environment

Stuart Walker

Principal Architect and Technology Advisor
Leapstone Systems, Inc.
Chairman, Architecture Working Group
Multiservice Forum

Introduction

Most service providers are now facing considerable growth challenges, with a requirement to increase revenue and margin in the face of intense competition and market decline in their traditional areas of operations. An initial strategy for many service providers against price and subscriber erosion is to offer triple- and quadruple-play service bundles, providing their subscribers with voice, video, data, and mobile services with the convenience of a single bill. However, in itself, the quad-play bundle may only provide a short-term strategy in terms of revenue growth and subscriber retention. Most service providers, irrespective of channel, can support the same set of bundled application, TV, and media offerings by assembling the right set of assets, either by acquiring them from third-party vendors or through partnerships. In this market, service differentiation becomes difficult, inevitably leading to (low) price differentiation with the associated knock-on effects in declining revenues.

By adopting an appropriate service delivery approach, service providers can break the cycle of continued price-based differentiation in an increasingly competitive marketplace by offering suites of services to address the thousands of individual service experience expectations and needs of subscribers. This paper provides an architectural overview of a service delivery environment that will enable service providers to achieve sustainable organic growth, increasing both average revenue per user (ARPU) and subscriber numbers while reducing subscriber churn.

Architectural Drivers

This section describes the key architectural drivers that a service delivery environment needs to fulfill a service provider's medium- and long-term needs.

Service Velocity and Service Agility

Two key enablers that a service delivery environment should facilitate are service velocity and service agility.

Service velocity is the ability to introduce a new service or product into the network quickly and cheaply, allowing the service provider to react quickly to market conditions. This is important for maximizing the first-to-market advantage for new services and minimizing any first-to-market advantage of competitors, allowing the service provider to act as a fast follower for any competitor services that prove successful.

Service agility allows services deployed by a service provider to be employed in a wide variety of end-subscriber propositions with minimum effort and expense on behalf of the service provider.

Taken together, these enablers can effectively change the landscape of the addressable market for enhanced service products. The two primary parameters for the revenue generated by any given services (aside from the price the subscriber pays) are the penetration (the number of subscribers using the service) and the duration (how long the product remains on the market).

Traditionally, because of the high cost associated with creating and delivering services, only high-penetration, long-duration services could be profitably deployed. Improvements in service velocity and service agility will enable the profitable delivery of services that may only exist for short periods of time, taking advantage of passing fashions and fads, or be targeted at smaller niches of the subscriber base. The reduced cost and time taken to introduce new service products to the market also allows operators to adopt a more entrepreneurial approach. Since the investment in launching a service is much lower than was traditionally the case, the operators can take greater risks in launching services and not be hampered by the long planning cycles associated with traditional service launch. The service providers will have the economic "freedom to fail" when introducing new service products.

Blended Services

As was briefly described earlier, a quad-play strategy of simply bundling services—providing multiple services to a subscriber with the convenience of a single bill—is unlikely to alleviate the downward pressure on revenue for most service providers.

The issues with simple service bundles are the likely ubiquitous nature of the offers, with most service providers providing the same set of services, leading to a price war among service providers. There is also something of an undermining of individual revenue streams for each of the bundled services in that end users will expect discounts for bundled offerings.

Service differentiation can be achieved, however, by offering blended services. A blended service is a service that makes use of two or more quad-play offerings (e.g., voice, video, data, mobile). A simple example of a blended service could be overlaying the caller ID of an incoming voice call on the television. Blended services are a result of service convergence; they can uniquely enhance the end-user experience, which can offset the erosion of service

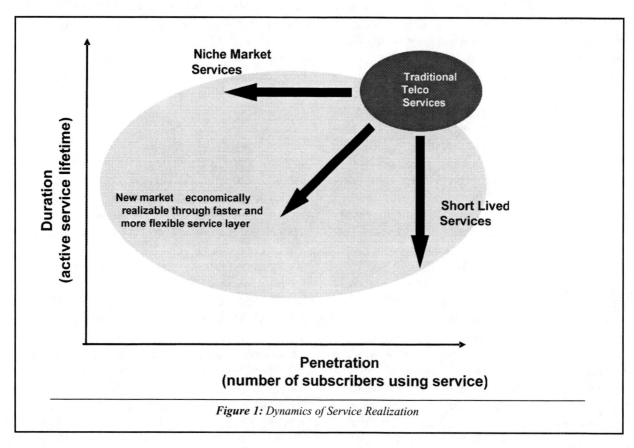

Figure 1: Dynamics of Service Realization

bundle price and provide a net increase in the ARPU.

An Architecture for Service Delivery

Figure 2 shows, schematically, a service delivery architecture that will meet the key business drivers of enabling the creation of blended services and achieving both service velocity and service agility.

At the base of the service delivery architecture "stack" are the constituent networks (voice, video, data, and mobile); the functions of these networks are exposed to the higher-level services as a set of capabilities or enablers. The enablers may take the form of relatively abstracted high-level functions (e.g., the Parlay and ParlayX application programming interfaces [APIs] defined by the Parlay group). The high-level abstracted functions can likely support a wide variety of application and service requirements, but some applications may require more direct access to specific functions of the underlying networks (some of the more obscure and esoteric aspects of the network protocols are there for a reason). To support this, a lower-level

capability set would provide protocol-level functions (e.g., the session initiation protocol [SIP] servlet definitions [JSR 116, JSR 289] providing application access to the SIP). As will be explained later, reusable service objects will allow the creation of higher-level, more abstract functions from lower-level ones.

The capabilities/enablers are used to create end-user services or, more correctly, reusable service objects. Creating a reusable service object is typically a programmatic task; a number of tools exist that perform this function, typically comprising a combination of a drag-and-drop graphical user interface (GUI) environment and a programming tool. Reusable service objects are described in more detail in the next section.

The service objects are grouped into products that form the offers made available to end subscribers. The product is an aggregation of service objects from a management perspective (allowing the aggregate to be managed atomically). The products are the managed objects visible to the operations support system (OSS) and business support system

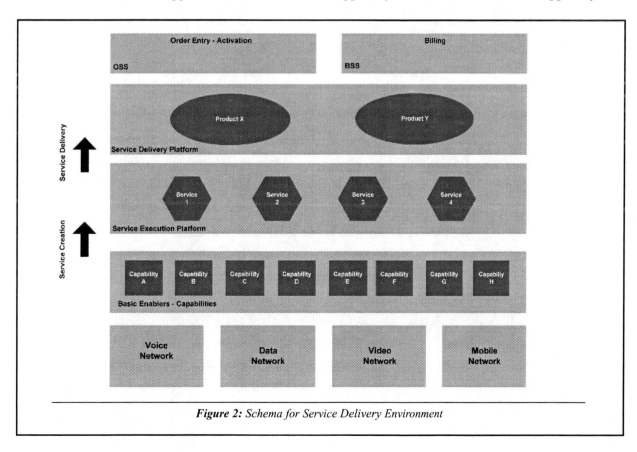

Figure 2: Schema for Service Delivery Environment

(BSS) functions; subscribers are activated and billed against products.

One important distinction of this architecture from earlier service delivery models is the inclusion of the OSS and BSS aspects of the service in the service delivery environment. Earlier models tended to separate the service creation function from the service integration function (interfacing the service to the necessary OSS and BSS) and focus on improving the efficiency of the service creation function. However, the integration task generally consumed far more effort and time than the service creation task, so improvements in the efficiency of the service creation function, welcome as they are, had a limited overall impact on reducing the time to market for new services.

Another important consideration is that a different approach and toolset will be used in the service creation function than in the service delivery function. Generally, it is expected that in such an architecture as that depicted in *Figure 2* there would be tens of capabilities and perhaps hundreds of services and thousands of products (service bundles). It is therefore necessary that the process to create a product by packing services together, as well as the process to create a capability or enabler, be considerably more lightweight that the process to create the services.

Reusable Service Objects

Figure 3 shows how reusable service objects are constructed in this service delivery architecture. The reusable service object will expose both an invocation interface (this could be a network-facing trigger event or a programmatic invocation interface, typically a Web service) and a management interface. The management interface exposes both provisioning/activation and billing functions for the reusable service object. The activation/provisioning interface is used to activate the service for a subscriber, which, in most cases, will involve the storage of subscriber-specific service information. Data repositories for subscriber- and service-specific data may themselves be provided as capabilities. The inclusion of the provisioning/activation func-

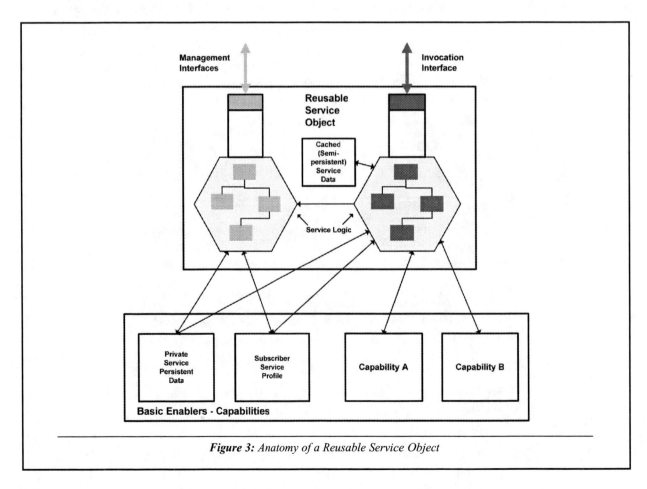

Figure 3: Anatomy of a Reusable Service Object

tions and associated logic within the reusable service object itself will provide a less brittle architecture than attempting to separate the run-time logic and the data it uses, since in the latter case the data schema needs to be maintained in multiple places.

The service logic of the service itself uses the underlying capabilities. For services that expose the invocation interface programmatically, then this interface could be used by other reusable service objects in the same way that they use capabilities. Over time the library of reusable service objects will increase, providing higher levels of functionality than the base capabilities enabling more rapid creation of services.

Service and Product Repositories

The end-to-end service creation and delivery process is shown in *Figure 4.*

Services (reusable service objects) are created in an application creation environment. These reusable service objects use the basic capabilities of the underlying networks and any already created reusable service objects. The definition of the created reusable service object is published in two logical internal directories. One for the invocation interface (if any), this directory is used by the application creation environment to discover the set of available reusable service objects. The management interfaces for the reusable service objects are stored in a separate logical directory; this directory is used by the service delivery platform (SDP) to discover the set of services that can be packaged into end-user products. The SDP provides a packaging function to package one or more service objects into end-user (orderable) products and publishes the product definitions (management interfaces of the service package as a whole plus any additional information) to a third directory, a directory of service packages. The directory of service packages is effectively the product catalog and is used by the order entry systems to determine the available products. The SDP is responsible for providing the translation between the ordered view of the product (the provisioning interface of the service bundle) and the activation interfaces of the individual services in the package.

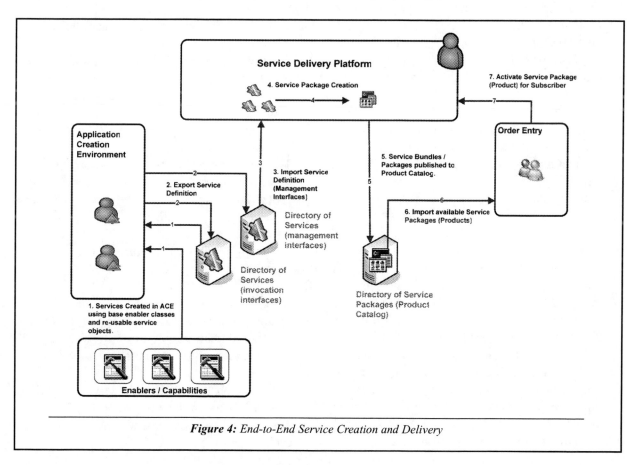

Figure 4: *End-to-End Service Creation and Delivery*

The steps shown in *Figure 4* are explained below.

1. A new service object is created in the application creation environment. The application creation environment discovers the available reusable service objects that can be used, along with the underlying capabilities, to create the new service objects.
2. Once created and suitably tested and approved, the new service object is published in the directory of services. There are two directories that may store information about the new service. The invocation interfaces are stored in a services directory, which is used by the application creation environment to discover the available reusable service objects. The management interfaces are stored in a services directory, used by the SDP to create the end-user propositions. Not all service objects will have entries in both service directories.
3. The SDP imports the services from the directory of services (management interface definitions). These form the set of service objects that can be used to create the end-user service bundles.
4. The SDP is used to create a new product (service bundle). The SDP would typically provide a high-level drag-and-drop or pick-list interface for this function.
5. The resulting product is published in a directory of service packages (product catalog). This directory will contain the management interfaces for the service package (the SDP holds the mapping of service packages to the individual services and exports the aggregate management functions).
6. The directory of service packages is imported into the order entry functions of the OSS as the list of products that are available to the end users.
7. The order entry system activates a subscriber against a service package via the aggregate provisioning interface exposed by the SDP (which the order entry system discovered from the directory of service packages). The SDP translates between the ordered view of the service package (the aggregate provisioning interface) and the individual activation interfaces of each service in the bundle (which it discovered from the directory of services). The SDP activates the individual service objects in the bundle.

Reusable Service Object Examples

Zip Change Notification

The zip change notification service object provides notifications for zip changes for either an individual address (representing a mobile endpoint) or a list of addresses. The address or list of addresses to monitor is passed to the service object through the invocation interface; no subscriber-level data needs be provisioned through the service object.

As shown in *Figure 5*, the zip change notification reusable service object uses two basic enablers/capabilities, location and group list. It is assumed in this example that these functions adhere to the ParlayX 2.0 specification, which means the location capability returns the coordinates of an endpoint in terms of longitude, latitude, and altitude (with a specified accuracy). The zip change notification reusable service object also uses another reusable service object in the form of location to zip; this reusable service object returns the zip code for a given latitude and longitude.

Figure 6 shows the basic logic flow of the zip change notification reusable service object, the major steps of which are explained as follows:

1. An invoking entity uses the reusable service object through its exposed Web service interface. The invoking entity passes the name of the address list that it wishes to monitor along with a correlation tag (the zip change notification can also accept a single address).
2. The zip change notification reusable service object retrieves the list of addresses from the group list enabler, using the query members request method (defined in the ParlayX 2.0 specification).
3. The list of addresses is returned, then for each address in the list:
4. The location enabler is invoked to obtain the current location of the address using the get location request method (defined in the Parlay X 2.0 specification).
5. The location information for the address is returned.
6. The location to zip reusable service object is invoked to obtain the zip code for the location.
7. The zip code is returned and stored in the local cache against the address.

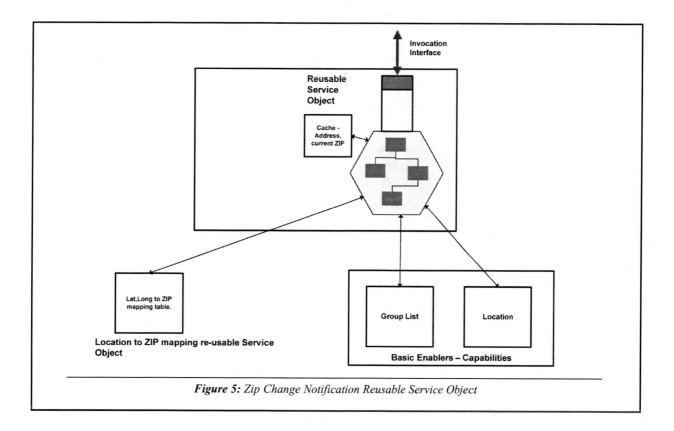

Figure 5: *Zip Change Notification Reusable Service Object*

8. The location enabler is instructed to provide notifications of any changes to the location of the address.

9. The enabler responds indicating that the address is being monitored for location changes.

10. The zip change notification reusable service object responds to the invoking entity with a list of addresses and zip codes of their current location.

11. An address changes location, and notification of the change is sent to the zip change notification reusable service object using the location notification request callback method (defined in the ParlayX 2.0 specification).

12. The zip change notification reusable service object responds to the notification.

13. The location to zip reusable service object is invoked to obtain the zip code for the returned location.

14. The zip code is returned.

15. If the zip code is different from that stored for the address, a notification is sent to the original invoking entity using the callback method (which forms part of the invocation interface description).

16. The invoking entity responds to the notification.

The next example shows how another end-user service can be built on top of the zip change notification.

In My Area

The in my area service provides notification to a subscriber, via a short message service (SMS), whenever a specified number of addresses from their address book are in the same zip code as the subscriber.

The in my area service object has no invocation interface (so it is not considered reusable, as other service objects cannot make use of it). Rather, the service is triggered by network events, in this case from the subscriber's own terminal registering with the network (this is described in more detail in the next section). The service object exports a management interface in order for it to be activated for a subscriber and to generate billing events for the sending of notifications.

The service object uses the zip change notification reusable service object described in the previous

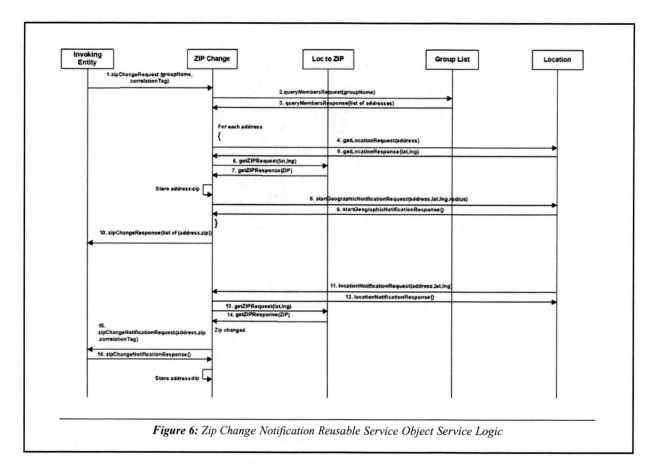

Figure 6: *Zip Change Notification Reusable Service Object Service Logic*

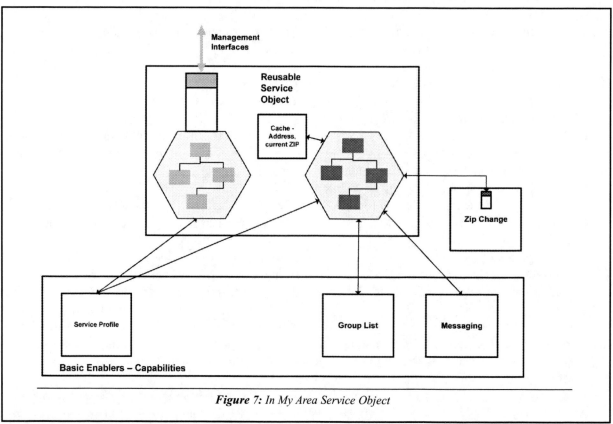

Figure 7: *In My Area Service Object*

section and three underlying capabilities, the group list, messaging capabilities (assumed to conform to the ParlayX 2.0 specification), and service profile, which provides a persistent data store for subscriber service information.

The service logic from the in my area service object is shown in *Figure 8* and explained below.

1. The in my area service object receives a notification event from the mobile network when the subscriber's device registers. In an IMS network this would be a SIP register message.
2. The in my area service object retrieves the subscriber service profile from the service profile enabler (in an IMS network this may be transparent data stored in the home subscriber server [HSS]).
3. The subscriber service data is returned, including the name of the address list to use and the parameter governing how many addresses must be in the subscriber's zip code to trigger a notification.
4. The zip change notification reusable service object is invoked with the subscriber's own address.
5. The zip change notification reusable service object returns the zip code of the subscriber's current location (and begins monitoring the subscriber's zip location).
6. The zip change notification reusable service object is invoked with the subscriber's address list retrieved from the service profile.
7. The zip change notification reusable service object returns the list of addresses and their current zip codes and begins monitoring the zip location of these addresses. There are currently insufficient addresses in the subscriber's zip code to trigger a notification message.
8. One of the addresses changes zip code, which is reported by the zip change notification reusable service object.
9. The in my area service object responds to the notification.
10. As there are now sufficient addresses in the subscriber's zip code, a notification is generated that includes the list of addresses in the subscriber's zip code. The notification is sent using the messaging enabler through the send SMS request method (defined in the ParlayX 2.0 specification).
11. The messaging enabler responds indicating the notification has been sent.
12. A billing event is generated by the in my area service object.
13. A response is received for the generated billing event.

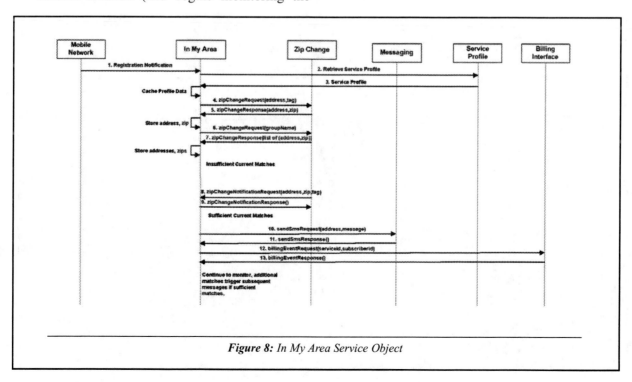

Figure 8: In My Area Service Object

Network Triggered Service Considerations

The in my area service object described in the previous section is invoked from a network event, the registration of the subscriber's device to the network. In an IMS network (which is the current target architecture for both mobile and fixed networks), this would have been a propagation of the SIP register message to the application platform hosting the in my area service logic. The HSS is an element in the IMS network that hosts the data used to trigger services from the network elements (primarily the serving call state control function [S–CSCF]). This data is termed the initial filter criteria (IFC) and is stored with the subscriber's service profile, identified by the SIP address of their terminal device. There is only one service profile that contains the triggering criteria, so it is necessary for the profile to be created at the SDP layer in the architecture described in this paper, since an individual service object effectively has no knowledge

of any other service objects that a subscriber may have in the service profile. This is also true at the product level, since a subscriber will often be activated against multiple products (service bundles).

To create the appropriate IFC for a subscriber's service profile, the SDP needs the actual triggering criteria for the service (e.g., voice mail triggering on destination busy or destination no answer) along with a set of static interaction rules. The static interaction rules will be used to determine both the ordering of services (priority) when they share the same triggering criteria and also to prevent incompatible services being added to the same IFC set.

In the example in *Figure 9*, the subscriber is activated against two products (X and Y). For the purposes of creating the IFC, this is "flattened" to the services that form part of that bundle, with any duplicates removed. Only four of the five resulting services are network-triggered services. The SDP com-

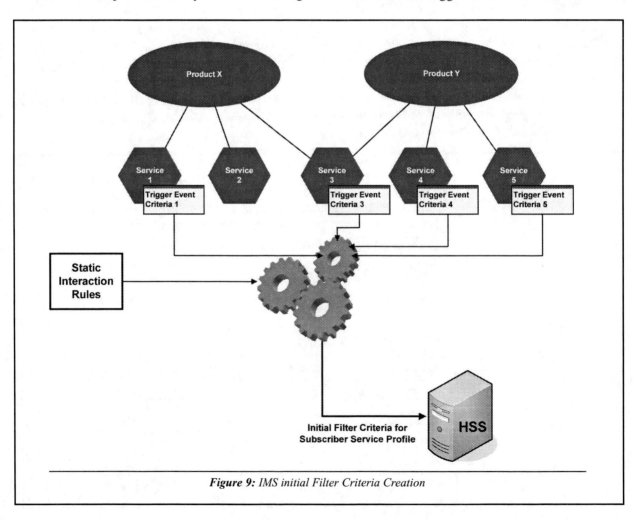

Figure 9: IMS initial Filter Criteria Creation

putes the IFC for the subscriber based on the triggering event criteria for each of the four services and any applicable static interaction rules. The resulting IFC is published to the HSS as part of the subscriber profile.

Each time a new product is activated or deactivated for the subscriber, the IFC may need to be recomputed if the change includes any network-initiated services.

Summary

The ability for service providers to offer new and compelling services that leverage all quad-play network assets will be key to increasing ARPU and retaining subscribers. The service architecture deployed by service providers needs to span all the quad-play networks, enabling service providers to offer differentiating blended services. The architecture also needs to enable service velocity to give service providers a rapid market reaction time, rapidly deploy and exploit new service ideas, and quickly reduce any competitor's first-to-market advantage.

This paper provided a high-level description of an architecture that meets these criteria and could form the cornerstone for any service provider's competitive quad-play strategy.

Acronym Guide

3G	third generation	DOCSIS	data over cable service interface specification
3GPP	third-generation partnership project	DoS	denial of service
ADM	add-drop multiplexer	DRM	digital rights management
ADSL	asymmetric digital subscriber line	DSCP	DiffServ code point
API	application program interface	DSL	digital subscriber line
APON	ATM–based passive optical network	DSLAM	digital subscriber line access multiplexer
ARPU	average revenue per user	DVR	digital video recorders
ATA	analog telephone adaptor	EDFA	Erbium-doped fiber amplifier
ATIS	Alliance for Telecommunications Industry Solutions	EFM	Ethernet in the first mile
		EMI	electromagnetic interference
ATM	asynchronous transfer mode	EMS	element management system
ATU–C	ADSL transceiver unit-central office	EPON	Ethernet passive optical network
		ER	edge router
ATU–R	ADSL transceiver unit-remote	ERS	Ethernet routing switch
BGCF	breakout gateway control function	ES	enhanced services
BGF	border gate function	ESU	Ethernet services unit
BLC	broadband loop carrier	ETSI	European Telecommunications Standards Institute
BPL	broadband over power line		
BPON	broadband passive optical network	FCC	Federal Communications Commission
B–RAS	broadband–remote access server	FDDI	fiber distributed data interface
BSA	broadband service aggregator	FDM	frequency division multiplexing
BSC	base station controller	FSAN	full-service access network
BSP	broadband service provider	FTTB	fiber-to-the-building OR fiber-to-the-business
BSS	business support system		
CATV	cable television	FTTC	fiber-to-the-curb
CCM	connection control manager	FTTCab	fiber-to-the-cabinet
CLASS	custom local-area signaling service	FTTH	fiber-to-the-home
		FTTN	fiber-to-the-neighborhood OR fiber-to-the-node
CMS	call-management server		
CO	central office	FTTP	fiber-to-the-premises
CoS	class of service	GbE	gigabit Ethernet
COTS	commercial off-the-shelf	GEPON	gigabit Ethernet passive optical network
CPE	customer-premises equipment		
CPL	call-processing language	GPON	gigabit passive optical network
CRM	customer-relationship management	GSM	Global System for Mobile Communications
CSA	carrier serving area	HDSL	high-bit-rate digital subscriber line
CSCF	call session control function		
CSP	communications service provider	HDTV	high-definition television
CSR	customer-service representative	HE	head end
CWDM	coarse wavelength division multiplexing	HFC	hybrid fiber coax
		HiperLAN	high-performance local-area network
DFS	dynamic frequency selection		
DHCP	dynamic host configuration protocol	HPNA	Home Phone Networking Alliance
		HSD	high-speed data
DLC	digital loop carrier	HSI	high-speed Internet

HSS	home subscriber server	MSO	multiple-system operators
HTML	hypertext markup language	MSPP	multiservice provisioning platform
HTTP	hypertext transfer protocol	MTA	media terminal adaptor
H–VPLS	hierarchical VPLS	MTTR	mean time to repair
IAD	integrated access device	MTU	multi-tenant unit
IBCF	interworking border gate control function	MVNO	mobile virtual network operator
		NAP	network access point
IEEE	Institute of Electrical and Electronics Engineers	NAP–R	network access point router
		NAS	network access server
IETF	Internet Engineering Task Force	NASS	network attachment subsystem
IGMP	Internet group management protocol	NAT	network address translation
		NAT–PT	network address translation–protocol translation
ILEC	incumbent local-exchange carrier		
IM	instant messaging	NCS	network call signaling
IMS	Internet protocol multimedia subsystem	NGN	next-generation network
		NIC	network interface card
IN	intelligent network	NID	network interface device
IP	Internet protocol	NLOS	near line-of-sight
IPSec	IP security	NMS	network management system
IPTV	Internet protocol television	NPVR	network personal video recorder
ISDN	integrated services digital network	NVoD	near video on demand
IS–IS	intermediate system to intermediate system	OADM	optical add-drop multiplexer
		OAM	operations, administration, and maintenance
ISO	International Organization for Standardization		
		ODN	optical distribution network
ISP	Internet service provider	OEM	original equipment manufacturer
ITU	International Telecommunications Union	OFDM	orthogonal frequency division multiplexing
ITU–T	International Telecommunication Union–Telecommunication Standardization Sector		
		OLT	optical line terminal
		ONT	optical network terminal
		ONU	optical network unit
L2TP	Layer-2 tunneling protocol	OSPF	open shortest path first
LAN	local-area network	OSS	operations support system
LOS	line-of-sight	P2MP	point-to-multipoint
LSP	label-switched path	P2P	point-to-point
MAC	media access control	P–CSCF	proxy call session control function
MAN	metro-area network	PDA	personal digital assistant
MEF	Metro Ethernet Forum	PDH	plesiochronous digital hierarchy
MEGACO	media gateway control	PESQ	perceptual evaluation of speech quality
MGCF	media gateway control function		
MGCP	media gateway control protocol	PHB	per-hop behavior
MOS	mean opinion score	PIM	personal information manager OR protocol-independent multicast
MPEG	Moving Pictures Experts Group		
MPLS	multiprotocol label switching	PLC	power-line communications
MRF	multimedia resource function	PON	passive optical network
MRFC	multimedia resource function controller	PoP	point of presence
		PoS	point of service
MRFP	multimedia resource function processor	POTS	plain old telephone service
		PPPoE	point-to-point protocol over Ethernet
MSC	mobile switching center		

PPTP	point-to-point tunneling protocol		STB	set-top box
PSN	packet-switched networks		TCO	total cost of ownership
PSTN	public switched telephone network		TCP	transmission control protocol
PTT	push-to-talk		TDM	time division multiplexing
PVC	permanent virtual circuit		TE–LSDB	traffic-engineering link-state database
PVR	personal video recorder			
QoS	quality of service		ToS	type of service
RACS	resource and admission control subsystem		TTS	text-to-speech
			UCE	universal call engine
RBAC	role-based access control		UDDI	universal description, discover, and integration
RBOC	regional Bell operating company			
RED	random early detection		UDP	user datagram protocol
RF	radio frequency		UNE–L	unbundled network element loop
RFI	radio-frequency interference		UNE–P	unbundled network element platform
ROADM	reconfigurable optical add-drop multiplexer		U–NII	unlicensed national information infrastructure
ROI	return on investment			
RT	remote terminal		URI	uniform resource identifier
RTCP	real-time transport control protocol		USB	universal serial bus
			VA–ISP	value-added Internet service provider
RTP	real-time transport protocol			
SBC	session border controller		VDSL	very-high–data-rate digital subscriber line
SCIM	service capability interaction manager			
			VHO	video hub office
SDH	synchronous digital hierarchy		VLAN	virtual local-area network
SDN	subscriber distribution network		VoD	video on demand
SDP	service delivery platform		VoIP	voice over IP
SDSL	symmetric digital subscriber line		VPN	virtual private network
SDTV	standard definition television		WAN	wide-area network
SHDSL	symmetric high-rate digital subscriber line		WDM	wavelength division multiplexing
			WFQ	weighted fair queuing
SHE	super head end		Wi-Fi	wireless fidelity
SIP	session initiation protocol		WiMAX	worldwide interoperability for microwave access
SLA	service-level agreement			
SLF	subscription locator function		WISP	wireless Internet service provider
SM	signaling manager		WLAN	wireless local-area network
SME	small-to-medium enterprise		WPAN	wireless personal-area network
SMP	service management platform		WPON	WDM passive optical network
SMS	short message service		WRED	weighted random early detection
SOA	service-oriented architecture		WRR	weighted round robin
SOAP	simple object access protocol		WSDL	Web Services description language
SOHO	small office/home office		WWAN	wireless wide-area network
SONET	synchronous optical network		XML	extensible markup language
SSL	secure sockets layer			